First Principles

Part I The Unknowable

by

Herbert Spencer

First Principles
Part I The Unknowable
by Herbert Spencer

ISBN: 978-93-61423-66-6

Published by

DOUBLE 9 BOOKS

2/13-B, Ansari Road
Daryaganj, New Delhi – 110002
info@double9books.com
www.double9books.com
Tel. 011-40042856

ABOUT THE AUTHOR

English polymath Herbert Spencer worked as a sociologist, anthropological, biologist, psychologist, and philosopher. The phrase "survival of the fittest" was first used by Spencer in Principles of Biology (1864), following his reading of Charles Darwin's 1859 book On the Origin of Species. Although the name primarily denotes natural selection, Spencer also embraced Lamarckism since he believed that evolution extends into the fields of sociology and ethics. Spencer created a comprehensive theory of evolution that included the progressive development of biological systems, the physical environment, human thought, culture, and society. He made contributions to many different fields as a polymath, such as politics, economics, anthropology, ethics, literature, astronomy, biology, sociology, and psychology. He attained great power throughout his lifetime, mostly in academic English-speaking circles. Although Spencer was "the single most famous European intellectual in the closing decades of the nineteenth century," his impact began to wane after 1900. Talcott Parsons questioned, "Who now reads Spencer?" in 1937. Spencer, the son of William George Spencer (often referred to as George), was born in Derby, England, on April 27, 1820.

CONTENTS

CHAPTER I
RELIGION AND SCIENCE 7

CHAPTER II
ULTIMATE RELIGIOUS IDEAS 23

CHAPTER III
ULTIMATE SCIENTIFIC IDEAS 39

CHAPTER IV
THE RELATIVITY OF ALL KNOWLEDGE 54

CHAPTER V
THE RECONCILIATION 76

CHAPTER I
RELIGION AND SCIENCE

§ 1. We too often forget that not only is there "a soul of goodness in things evil," but very generally also, a soul of truth in things erroneous. While many admit the abstract probability that a falsity has usually a nucleus of reality, few bear this abstract probability in mind, when passing judgment on the opinions of others. A belief that is finally proved to be grossly at variance with fact, is cast aside with indignation or contempt; and in the heat of antagonism scarcely any one inquires what there was in this belief which commended it to men's minds. Yet there must have been something. And there is reason to suspect that this something was its correspondence with certain of their experiences: an extremely limited or vague correspondence perhaps; but still, a correspondence. Even the absurdest report may in nearly every instance be traced to an actual occurrence; and had there been no such actual occurrence, this preposterous misrepresentation of it would never have existed. Though the distorted or magnified image transmitted to us through the refracting medium of rumour, is utterly unlike the reality; yet in the absence of the reality there would have been no distorted or magnified image. And thus it is with human beliefs in general. Entirely wrong as they may appear, the implication is that they germinated out of actual experiences—originally contained, and perhaps still contain, some small amount of verity.

More especially may we safely assume this, in the case of beliefs that have long existed and are widely diffused; and most of all so, in the case of beliefs that are perennial and nearly or quite universal. The presumption that any current opinion is not wholly false, gains in strength according to the number of its adherents. Admitting, as we must, that life is impossible unless through a certain agreement between internal convictions and external circumstances; admitting therefore that the probabilities are always in favour of the truth, or at least the partial truth, of a conviction; we must admit that the convictions entertained by many minds in common are the most likely to have some foundation. The elimination of individual errors of thought, must give to the resulting judgment a certain additional value. It may indeed be urged that many widely-spread beliefs are received on authority; that those entertaining them make no attempts at verification; and hence it may be inferred that the multitude of adherents adds but little to the probability of a belief. But this is not true. For a belief which gains extensive reception without critical examination, is thereby proved to have a general congruity with the various other beliefs of those who receive it; and in so far as these various other beliefs are based upon personal observation and judgment, they give an indirect warrant to one with which they harmonize. It may be that this warrant is of small value; but still it is of some value.

Could we reach definite views on this matter, they would be extremely useful to us. It is important that we should, if possible, form something like a general theory of current opinions; so that we may neither over-estimate nor under-estimate their worth. Arriving at correct judgments on disputed questions, much depends on the attitude of mind we preserve while listening to, or taking part in, the controversy; and for the preservation of a right attitude, it is needful that we should learn how true, and yet how untrue, are average human beliefs. On the one hand, we must keep free from that bias in favour of received ideas which expresses itself in such dogmas as "What every one says must be true," or "The voice of the people is the voice of God." On the other hand, the fact disclosed by a survey of the past, that majorities have usually been wrong, must not blind us to the complementary fact, that majorities have usually not been *entirely* wrong. And the avoidance of these extremes being a prerequisite to catholic thinking, we shall do well to provide ourselves with a safe-guard against them, by making a valuation of opinions in the abstract. To this end we must contemplate the kind of relation that ordinarily subsists between opinions and facts. Let us do so with one of those beliefs which under various forms has prevailed among all nations in all times.

§ 2. The earliest traditions represent rulers as gods or demigods. By their subjects, primitive kings were regarded as superhuman in origin, and superhuman in power. They possessed divine titles; received obeisances like those made before the altars of deities; and were in some cases actually worshipped. If there needs proof that the divine and half-divine characters originally ascribed to monarchs were ascribed literally, we have it in the fact that there are still existing savage races, among whom it is held that the chiefs and their kindred are of celestial origin, or, as elsewhere, that only the chiefs have souls. And of course along with beliefs of this kind, there existed a belief in the unlimited power of the ruler over his subjects—an absolute possession of them, extending even to the taking of their lives at will: as even still in Fiji, where a victim stands unbound to be killed at the word of his chief; himself declaring, "whatever the king says must be done."

In times and among races somewhat less barbarous, we find these beliefs a little modified. The monarch, instead of being literally thought god or demigod, is conceived to be a man having divine authority, with perhaps more or less of divine nature. He retains however, as in the East to the present day, titles expressing his heavenly descent or relationships; and is still saluted in forms and words as humble as those addressed to the Deity. While the lives and properties of his people, if not practically so completely at his mercy, are still in theory supposed to be his.

Later in the progress of civilization, as during the middle ages in Europe, the current opinions respecting the relationship of rulers and ruled are further changed. For the theory of divine origin, there is substituted that of divine right. No longer god or demigod, or even god-descended, the king is now regarded as simply God's vice-gerent. The obeisances made to him are not so extreme in their humility; and his sacred titles lose much of their meaning. Moreover his authority ceases to be unlimited. Subjects deny his right to dispose at will of their lives and properties; and yield allegiance only in the shape of obedience to his commands.

With advancing political opinion has come still greater restriction of imperial power. Belief in the supernatural character of the ruler, long ago repudiated by ourselves for example, has left behind it nothing more than the popular tendency to ascribe unusual goodness, wisdom, and beauty to the monarch. Loyalty, which originally meant implicit submission to the king's will, now means a merely nominal profession of subordination, and the fulfilment of certain forms of respect. Our political practice, and our political theory, alike utterly reject those regal prerogatives which once passed unquestioned. By deposing some, and putting others in their places, we have not only denied the divine rights of certain men to rule; but we have denied that they have any rights beyond those originating in the assent of the nation. Though our forms of speech and our state-documents still assert the subjection of the citizens to the ruler, our actual beliefs and our daily proceedings implicitly assert the contrary. We obey no laws save those of our own making. We have entirely divested the monarch of legislative power; and should immediately rebel against his or her exercise of such power, even in matters of the smallest concern. In brief, the aboriginal doctrine is all but extinct among us.

Nor has the rejection of primitive political beliefs, resulted only in transferring the authority of an autocrat to a representative body. The views entertained respecting governments in general, of whatever form, are now widely different from those once entertained. Whether popular or despotic, governments were in ancient times supposed to have unlimited authority over their subjects. Individuals existed for the benefit of the State; not the

State for the benefit of individuals. In our days, however, not only has the national will been in many cases substituted for the will of the king; but the exercise of this national will has been restricted to a much smaller sphere. In England, for instance, though there has been established no definite theory setting bounds to governmental authority; yet, in practice, sundry bounds have been set to it which are tacitly recognized by all. There is no organic law formally declaring that the legislature may not freely dispose of the citizens' lives, as early kings did when they sacrificed hecatombs of victims; but were it possible for our legislature to attempt such a thing, its own destruction would be the consequence, rather than the destruction of citizens. How entirely we have established the personal liberties of the subject against the invasions of State-power, would be quickly demonstrated, were it proposed by Act of Parliament forcibly to take possession of the nation, or of any class, and turn its services to public ends; as the services of the people were turned by primitive rulers. And should any statesman suggest a re-distribution of property such as was sometimes made in ancient democratic communities, he would be met by a thousand-tongued denial of imperial power over individual possessions. Not only in our day have these fundamental claims of the citizen been thus made good against the State, but sundry minor claims likewise. Ages ago, laws regulating dress and mode of living fell into disuse; and any attempt to revive them would prove the current opinion to be, that such matters lie beyond the sphere of legal control. For some centuries we have been asserting in practice, and have now established in theory, the right of every man to choose his own religious beliefs, instead of receiving such beliefs on State-authority. Within the last few generations we have inaugurated complete liberty of speech, in spite of all legislative attempts to suppress or limit it. And still more recently we have claimed and finally obtained under a few exceptional restrictions, freedom to trade with whomsoever we please. Thus our political beliefs are widely different from ancient ones, not only as to the proper depositary of power to be exercised over a nation, but also as to the extent of that power.

Not even here has the change ended. Besides the average opinions which we have just described as current among ourselves, there exists a less widely-diffused opinion going still further in the same direction. There are to be found men who contend that the sphere of government should be narrowed even more than it is in England. The modern doctrine that the State exists for the benefit of citizens, which has now in a great measure supplanted the ancient doctrine that the citizens exist for the benefit of the State, they would push to its logical results. They hold that the freedom of the individual, limited only by the like freedom of other individuals, is sacred; and that the legislature cannot equitably put further restrictions upon it, either by forbidding any actions which the law of equal freedom permits, or taking away any property save that required to pay the cost of enforcing this law itself. They assert that the sole function of the State is the protection of persons against each other, and against a foreign foe. They urge that as, throughout civilization, the manifest tendency has been continually to extend the liberties of the subject, and restrict the functions of the State, there is reason to believe that the ultimate political condition must be one in which personal freedom is the greatest possible and governmental power the least possible: that, namely, in which the freedom of each has no limit but the like freedom of all; while the sole governmental duty is the maintenance of this limit.

Here then in different times and places we find concerning the origin, authority, and functions of government, a great variety of opinions—opinions of which the leading genera above indicated subdivide into countless species. What now must be said about the truth or falsity of these opinions? Save among a few barbarous tribes the notion that a monarch is a god or demigod is regarded throughout the world as an absurdity almost passing the bounds of human credulity. In but few places does there survive a vague notion that the ruler possesses any supernatural attributes. Most civilized communities, which still admit the divine right of governments, have long since repudiated the divine right of kings. Elsewhere the belief that there is anything sacred in legislative regulations is dying out: laws are coming to be considered as conventional only. While the extreme school holds that governments have neither intrinsic authority, nor can have authority given to them by convention; but can possess authority only as the administrators of those moral principles deducible from the conditions essential to social life. Of these various beliefs, with their innumerable modifications, must we then say that some one alone is wholly right and all the rest wholly wrong; or must we say that each of them contains truth more or less completely disguised by errors? The latter alternative is the one which analysis will force upon us. Ridiculous as they may severally appear to those not educated under them, every one of these doctrines has for its vital element the recognition of an unquestionable fact. Directly or by implication, each of them insists on a certain subordination of individual actions to social requirements. There are wide differences as to the power to which this subordination is due; there are wide differences as to the motive for this subordination; there are wide differences as to its extent; but that there must be *some* subordination all are agreed. From the oldest and rudest idea of allegiance, down to the most advanced political theory of our own day, there is on this point complete unanimity. Though, between the savage who conceives his life and property to be at the absolute disposal of his chief, and the anarchist who denies the right of any government, autocratic or democratic, to trench upon his individual freedom, there seems at first sight an entire and irreconcilable antagonism; yet ultimate analysis discloses in them this fundamental community of opinion; that there are limits which individual actions may not transgress—limits which the one regards as originating in the king's will, and which the other regards as deducible from the equal claims of fellow-citizens.

It may perhaps at first sight seem that we here reach a very unimportant conclusion; namely, that a certain tacit assumption is equally implied in all these conflicting political creeds—an assumption which is indeed of self-evident validity. The question, however, is not the value or novelty of the particular truth in this case arrived at. My aim has been to exhibit the more general truth, which we are apt to overlook, that between the most opposite beliefs there is usually something in common,—something taken for granted by each; and that this something, if not to be set down as an unquestionable verity, may yet be considered to have the highest degree of probability. A postulate which, like the one above instanced, is not consciously asserted but unconsciously involved; and which is unconsciously involved not by one man or body of men, but by numerous bodies of men who diverge in countless ways and degrees in the rest of their beliefs; has a warrant far transcending any that can be usually shown. And when, as in this case, the postulate is abstract—is not based on some one concrete experience common to all mankind, but implies an induction from a great variety of experiences, we may say that it ranks next in certainty to the postulates of exact science.

Do we not thus arrive at a generalization which may habitually guide us when seeking for the soul of truth in things erroneous? While the foregoing illustration brings clearly home the fact, that in opinions seeming to be absolutely and supremely wrong something right is yet to be found; it also indicates the method we should pursue in seeking the something right. This method is to compare all opinions of the same genus; to set aside as more or less discrediting one another those various special and concrete elements in which such opinions disagree; to observe what remains after the discordant constituents have been eliminated; and to find for this remaining constituent that abstract expression which holds true throughout its divergent modifications.

§3. A candid acceptance of this general principle and an adoption of the course it indicates, will greatly aid us in dealing with those chronic antagonisms by which men are divided. Applying it not only to current ideas with which we are personally unconcerned, but also to our own ideas and those of our opponents, we shall be led to form far more correct judgments. We shall be ever ready to suspect that the convictions we entertain are not wholly right, and that the adverse convictions are not wholly wrong. On the one hand we shall not, in common with the great mass of the unthinking, let our beliefs be determined by the mere accident of birth in a particular age on a particular part of the Earth's surface; and, on the other hand, we shall be saved from that error of entire and contemptuous negation, which is fallen into by most who take up an attitude of independent criticism.

Of all antagonisms of belief, the oldest, the widest, the most profound and the most important, is that between Religion and Science. It commenced when the recognition of the simplest uniformities in surrounding things, set a limit to the previously universal fetishism. It shows itself everywhere throughout the domain of human knowledge: affecting men's interpretations alike of the simplest mechanical accidents and of the most complicated events in the histories of nations. It has its roots deep down in the diverse habits of thought of different orders of minds. And the conflicting conceptions of nature and life which these diverse habits of thought severally generate, influence for good or ill the tone of feeling and the daily conduct.

An unceasing battle of opinion like this which has been carried on throughout all ages under the banners of Religion and Science, has of course generated an animosity fatal to a just estimate of either party by the other. On a larger scale, and more intensely than any other controversy, has it illustrated that perennially significant fable concerning the knights who fought about the colour of a shield of which neither looked at more than one face. Each combatant seeing clearly his own aspect of the question, has charged his opponent with stupidity or dishonesty in not seeing the same aspect of it; while each has wanted the candour to go over to his opponent's side and find out how it was that he saw everything so differently.

Happily the times display an increasing catholicity of feeling, which we shall do well in carrying as far as our natures permit. In proportion as we love truth more and victory less, we shall become anxious to know what it is which leads our opponents to think as they do. We shall begin to suspect that the pertinacity of belief exhibited by them must result from a perception of something we have not perceived. And we shall aim to supplement the portion of truth we have found with the portion found by them. Making a more rational estimate of human authority, we shall avoid alike the extremes of undue submission and undue rebellion—shall not regard some men's judgments as wholly good and others as wholly bad; but shall rather lean to the more defensible position that none are completely right and none are completely wrong.

Preserving, as far as may be, this impartial attitude, let us then contemplate the two sides of this great controversy. Keeping guard against the bias of education and shutting out the whisperings of sectarian feeling, let us consider what are the *à priori* probabilities in favour of each party.

§4. When duly realized, the general principle above illustrated must lead us to anticipate that the diverse forms of religious belief which have existed and which still exist, have all a basis in some ultimate fact. Judging by analogy the implication is, not that any one of them is altogether right; but that in each there is something right more or less disguised by other things wrong. It may be that the soul of truth contained in erroneous creeds is very unlike most, if not all, of its several embodiments; and indeed, if, as we have good reason to expect, it is much more abstract than any of them, its unlikeness necessarily follows. But however different from its concrete expressions, some essential verity must be looked for. To suppose that these multiform conceptions should be one and all *absolutely* groundless, discredits too profoundly that average human intelligence from which all our individual intelligences are inherited.

This most general reason we shall find enforced by other more special ones. To the presumption that a number of diverse beliefs of the same class have some common foundation in fact, must in this case be added a further presumption derived from the omnipresence of the beliefs. Religious ideas of one kind or other are almost if not quite universal. Even should it be true, as alleged, that there exist tribes of men who have nothing approaching to a theory of creation—even should it be true that only when a certain phase of intelligence is reached do the most rudimentary of such theories make their appearance; the implication is practically the same. Grant that among all races who have passed a certain stage of intellectual development there are found vague notions concerning the origin and hidden nature of surrounding things; and there arises the inference that such notions are necessary products of progressing intelligence. Their endless variety serves but to strengthen this conclusion: showing as it does a more or less independent genesis—showing how, in different places and times, like conditions have led to similar trains of thought, ending in analogous results. That these countless different, and yet allied, phenomena presented by all religions are accidental or factitious, is an untenable supposition. A candid examination of the evidence quite negatives the doctrine maintained by some, that creeds are priestly inventions. Even as a mere question of probabilities it cannot rationally be concluded that in every society, past and present, savage and civilized, certain members of the community have combined to delude the rest, in ways so analogous. To any who may allege that some primitive fiction was devised by some primitive priesthood, before yet mankind had diverged from a common centre, a reply is furnished by philology; for philology proves the dispersion of mankind to have commenced before there existed a language sufficiently organized to express religious ideas. Moreover, were it otherwise tenable, the hypothesis of artificial origin fails to account for the facts. It does not explain why, under all changes of form, certain elements of religious belief remain constant. It does not show us how it happens that while adverse criticism has from age to age gone on destroying particular theological dogmas, it has not destroyed the fundamental conception underlying these dogmas. It leaves us without any solution of the striking circumstance that when, from the absurdities and corruptions accumulated around them, national creeds have fallen into general discredit, ending in indifferentism or positive denial, there has always by and by arisen a re-assertion of them: if not the same in form, still the same in essence. Thus the universality of religious ideas, their independent evolution among different primitive races, and their great vitality, unite in showing that their source must be deep-seated instead of superficial. In other words, we are obliged to admit that if not supernaturally derived as the majority contend, they must be derived out of human experiences, slowly accumulated and organized.

Should it be asserted that religious ideas are products of the religious sentiment, which, to satisfy itself, prompts imaginations that it afterwards projects into the external world, and by and by mistakes for realities; the problem is not solved, but only removed further back. Whether the wish is father to the thought, or whether sentiment and idea have a common genesis, there equally arises the question—Whence comes the sentiment? That it is a constituent in man's nature is implied by the hypothesis; and cannot indeed be denied by those who prefer other hypotheses. And if the religious sentiment, displayed habitually by the majority of mankind, and occasionally aroused even in those seemingly devoid of it, must be classed among human emotions, we cannot rationally ignore it. We are bound to ask its origin and its function. Here is an attribute which, to say the least, has had an enormous influence—which has played a conspicuous part throughout the entire past as far back as history records, and is at present the life of numerous institutions, the stimulus to perpetual controversies, and the prompter of countless daily actions. Any Theory of Things which takes no account of this attribute, must, then, be extremely defective. If with no other view, still as a question in philosophy, we are called on to say what this attribute means; and we cannot decline the task without confessing our philosophy to be incompetent.

Two suppositions only are open to us: the one that the feeling which responds to religious ideas resulted, along with all other human faculties, from an act of special creation; the other that it, in common with the rest, arose by a process of evolution. If we adopt the first of these alternatives, universally accepted by our ancestors and by the immense majority of our contemporaries, the matter is at once settled: man is directly endowed with the religious feeling by a creator; and to that creator it designedly responds. If we adopt the second alternative, then we are met by the questions—What are the circumstances to which the genesis of the religious feeling is due? and—What is its office? We are bound to entertain these questions; and we are bound to find answers to them. Considering all faculties, as we must on this supposition, to result from accumulated modifications caused by the intercourse of the organism with its environment, we are obliged to admit that there exist in the environment certain phenomena or conditions which have determined the growth of the feeling in question; and so are obliged to admit that it is as normal as any other faculty. Add to which that as, on the hypothesis of a development of lower forms into higher, the end towards which the progressive changes directly or indirectly tend, must be adaptation to the requirements of existence; we are also forced to infer that this feeling is in some way conducive to human welfare. Thus both alternatives contain the same ultimate implication. We must conclude that the religious sentiment is either directly created, or is created by the slow action of natural causes; and whichever of these conclusions we adopt, requires us to treat the religious sentiment with respect.

One other consideration should not be overlooked—a consideration which students of Science more especially need to have pointed out. Occupied as such are with established truths, and accustomed to regard things not already known as things to be hereafter discovered, they are liable to forget that information, however extensive it may become, can never satisfy inquiry. Positive knowledge does not, and never can, fill the whole region of possible thought. At the uttermost reach of discovery there arises, and must ever arise, the question—What lies beyond? As it is impossible to think of a limit to space so as to exclude the idea of space lying outside that limit; so we cannot conceive of any explanation profound enough to exclude the question—What is the explanation of that explanation? Regarding Science as a gradually increasing sphere, we may say that every addition to its surface does but bring it into wider contact with surrounding nescience. There must ever remain therefore two antithetical modes of mental action. Throughout all future time, as now, the human mind may occupy itself, not only with ascertained phenomena and their relations, but also with that unascertained something which phenomena and their relations imply. Hence if knowledge cannot monopolize consciousness—if it must always continue possible for the mind to dwell upon that which transcends knowledge; then there can never cease to be a place for something of the nature of Religion; since Religion under all its forms is distinguished from everything else in this, that its subject matter is that which passes the sphere of experience.

Thus, however untenable may be any or all the existing religious creeds, however gross the absurdities associated with them, however irrational the arguments set forth in their defence, we must not ignore the verity which in all likelihood lies hidden within them. The general probability that widely-spread beliefs are not absolutely baseless, is in this case enforced by a further probability due to the omnipresence of the beliefs. In the existence of a religious sentiment, whatever be its origin, we have a second evidence of great significance. And as in that nescience which must ever remain the antithesis to science, there is a sphere for the exercise of this sentiment, we find a third general fact of like implication. We may be sure therefore that religions, though even none of them be actually true, are yet all adumbrations of a truth.

§ 5. As, to the religious, it will seem absurd to set forth any justification for Religion; so, to the scientific, will it seem absurd to defend Science. Yet to do the last is certainly as needful as to do the first. If there exists a class who, in contempt of its follies and disgust at its corruptions, have contracted towards Religion a repugnance which makes them overlook the fundamental verity contained in it; so, too, is there a class offended to such a degree by the destructive criticisms men of science make on the religious tenets they regard as essential, that they have acquired a strong prejudice against Science in general. They are not prepared with any avowed reasons for their dislike. They have simply a remembrance of the rude shakes which Science has given to many of their cherished convictions, and a suspicion that it may perhaps eventually uproot all they regard as sacred; and hence it produces in them a certain inarticulate dread.

What is Science? To see the absurdity of the prejudice against it, we need only remark that Science is simply a higher development of common knowledge; and that if Science is repudiated, all knowledge must be repudiated along with it. The extremest bigot will not suspect any harm in the observation that the sun rises earlier and sets later in the summer than in the winter; but will rather consider such an observation as a useful aid in fulfilling the duties of life. Well, Astronomy is an organized body of similar observations, made with greater nicety, extended to a larger number of objects, and so analyzed as to disclose the real arrangements of the heavens, and to dispel our false conceptions of them. That iron will rust in water, that wood will burn, that long kept viands become putrid, the most timid sectarian will teach without alarm, as things useful to be known. But these are chemical truths: Chemistry is a systematized collection of such facts, ascertained with precision, and so classified and generalized as to enable us to say with certainty, concerning each simple or compound substance, what change will occur in it under given conditions. And thus is it with all the sciences. They severally germinate out of the experiences of daily life; insensibly as they grow they draw in remoter, more numerous, and more complex experiences; and among these, they ascertain laws of dependence like those which make up our knowledge of the most familiar objects. Nowhere is it possible to draw a line and say—here Science begins. And as it is the function of common observation to serve for the guidance of conduct; so, too, is the guidance of conduct the office of the most recondite and abstract inquiries of Science. Through the countless industrial processes and the various modes of locomotion which it has given to us, Physics regulates more completely our social life than does his acquaintance with the properties of surrounding bodies regulate the life of the savage. Anatomy and Physiology, through their effects on the practice of medicine and hygiene, modify our actions almost as much as does our acquaintance with the evils and benefits which common environing agencies may produce on our bodies. All Science is prevision; and all prevision ultimately aids us in greater or less degree to achieve the good and avoid the bad. As certainly as the perception of an object lying in our path warns us against stumbling over it; so certainly do those more complicated and subtle perceptions which constitute Science, warn us against stumbling over intervening obstacles in the pursuit of our distant ends. Thus being one in origin and function, the simplest forms of cognition and the most complex must be dealt with alike. We are bound in consistency to receive the widest knowledge which our faculties can reach, or to reject along with it that narrow knowledge possessed by all. There is no logical alternative between accepting our intelligence in its entirety, or repudiating even that lowest intelligence which we possess in common with brutes.

To ask the question which more immediately concerns our argument—whether Science is substantially true?—is much like asking whether the sun gives light. And it is because they are conscious how undeniably valid are most of its propositions, that the theological party regard Science with so much secret alarm. They know that during the two thousand years of its growth, some of its larger divisions—mathematics, physics, astronomy—have been subject to the rigorous criticism of successive generations; and have notwithstanding become ever more firmly established. They know that, unlike many of their own doctrines, which were once universally received but have age by age been more frequently called in question, the doctrines of Science, at first confined to a few scattered inquirers, have been slowly growing into general acceptance, and are now in great part admitted as beyond dispute. They know that men of science throughout the world subject each other's results to the most searching examination; and that error is mercilessly exposed and rejected as soon as discovered. And, finally, they know that still more conclusive testimony is to be found in the daily verification of scientific predictions, and in the never-ceasing triumphs of those arts which Science guides.

To regard with alienation that which has such high credentials is a folly. Though in the tone which many of the scientific adopt towards them, the defenders of Religion may find some excuse for this alienation; yet the excuse is a very insufficient one. On the side of Science, as on their own side, they must admit that short-comings in the advocates do not tell essentially against that which is advocated. Science must be judged by itself: and so judged, only the most perverted intellect can fail to see that it is worthy of all reverence. Be there or be there not any other revelation, we have a veritable revelation in Science—a continuous disclosure, through the intelligence with which we are endowed, of the established order of the Universe. This disclosure it is the duty of every one to verify as far as in him lies; and having verified, to receive with all humility.

§6. On both sides of this great controversy, then, truth must exist. An unbiassed consideration of its general aspects forces us to conclude that Religion, everywhere present as a weft running through the warp of human history, expresses some eternal fact; while it is almost a truism to say of Science that it is an organised mass of facts, ever growing, and ever being more completely purified from errors. And if both have bases in the reality of things, then between them there must be a fundamental harmony. It is an incredible hypothesis that there are two orders of truth, in absolute and everlasting opposition. Only on some Manichean theory, which among ourselves no one dares openly avow however much his beliefs may be tainted by it, is such a supposition even conceivable. That Religion is divine and Science diabolical, is a proposition which, though implied in many a clerical declamation, not the most vehement fanatic can bring himself distinctly to assert. And whoever does not assert this, must admit that under their seeming antagonism lies hidden an entire agreement.

Each side, therefore, has to recognize the claims of the other as standing for truths that are not to be ignored. He who contemplates the Universe from the religious point of view, must learn to see that this which we call Science is one constituent of the great whole; and as such ought to be regarded with a sentiment like that which the remainder excites. While he who contemplates the universe from the scientific point of view, must learn to see that this which we call Religion is similarly a constituent of the great whole; and being such, must be treated as a subject of science with no more prejudice than any other reality. It behoves each party to strive to understand the other, with the conviction that the other has something worthy to be understood; and with the conviction that when mutually recognized this something will be the basis of a complete reconciliation.

How to find this something—how to reconcile them, thus becomes the problem which we should perseveringly try to solve. Not to reconcile them in any makeshift way—not to find one of those compromises we hear from time to time proposed, which their proposers must secretly feel are artificial and temporary; but to arrive at the terms of a real and permanent peace between them. The thing we have to seek out, is that ultimate truth which both will avow with absolute sincerity—with not the remotest mental reservation. There shall be no concession—no yielding on either side of something that will by and by be reasserted; but the common ground on which they meet shall be one which each will maintain for itself. We have to discover some fundamental verity which Religion will assert, with all possible emphasis, in the absence of Science; and which Science, with all possible emphasis, will assert in the absence of Religion—some fundamental verity in the defence of which each will find the other its ally.

Or, changing the point of view, our aim must be to co-ordinate the seemingly opposed convictions which Religion and Science embody. From the coalescence of antagonist ideas, each containing its portion of truth, there always arises a higher development. As in Geology when the igneous and aqueous hypotheses were united, a rapid advance took place; as in Biology we are beginning to progress through the fusion of the doctrine of types with the doctrine of adaptations; as in Psychology the arrested growth recommences now that the disciples of Kant and those of Locke have both their views recognized in the theory that organized experiences produce forms of thought; as in Sociology, now that it is beginning to assume a positive character, we find a recognition of both the party of progress and the party of order, as each holding a truth which forms a needful complement to that held by the other; so must it be on a grander scale with Religion and Science. Here too we must look for a conception which combines the conclusions of both; and here too we may expect important results from their combination. To understand how Science and Religion express opposite sides of the same fact—the one its near or visible side, and the other its remote or invisible side—this it is which we must attempt; and to achieve this must profoundly modify our general Theory of Things.

Already in the foregoing pages the method of seeking such a reconciliation has been vaguely foreshadowed. Before proceeding further, however, it will be well to treat the question of method more definitely. To find that truth in which Religion and Science coalesce, we must know in what direction to look for it, and what kind of truth it is likely to be.

§ 7. We have found *à priori* reason for believing that in all religions, even the rudest, there lies hidden a fundamental verity. We have inferred that this fundamental verity is that element common to all religions, which remains after their discordant peculiarities have been mutually cancelled. And we have further inferred that this element is almost certain to be more abstract than any current religious doctrine. Now it is manifest that only in some highly abstract proposition, can Religion and Science find a common ground. Neither such dogmas as those of the trinitarian and unitarian, nor any such idea as that of propitiation, common though it may be to all religions, can serve as the desired basis of agreement; for Science cannot recognize beliefs like these: they lie beyond its sphere. Hence we see not only that, judging by analogy, the essential truth contained in Religion is that most abstract element pervading all its forms; but also that this most abstract element is the only one in which Religion is likely to agree with Science.

Similarly if we begin at the other end, and inquire what scientific truth can unite Science and Religion. It is at once manifest that Religion can take no cognizance of special scientific doctrines; any more than Science can take cognizance of special religious doctrines. The truth which Science asserts and Religion indorses cannot be one furnished by mathematics; nor can it be a physical truth; nor can it be a truth in chemistry: it cannot be a truth belonging to any particular science. No generalization of the phenomena of space, of time, of matter, or of force, can become a Religious conception. Such a conception, if it anywhere exists in Science, must be more general than any of these—must be one underlying all of them. If there be a fact which Science recognizes in common with Religion, it must be that fact from which the several branches of Science diverge, as from their common root.

Assuming then, that since these two great realities are constituents of the same mind, and respond to different aspects of the same Universe, there must be a fundamental harmony between them; we see good reason to conclude that the most abstract truth contained in Religion and the most abstract truth contained in Science must be the one in which the two coalesce. The largest fact to be found within our mental range must be the one of which we are in search. Uniting these positive and negative poles of human thought, it must be the ultimate fact in our intelligence.

§ 8. Before proceeding in the search for this common datum let me bespeak a little patience. The next three chapters, setting out from different points and converging to the same conclusion, will be comparatively unattractive. Students of philosophy will find in them much that is more or less familiar; and to most of those who are unacquainted with the literature of modern metaphysics, they may prove somewhat difficult to follow.

Our argument however cannot dispense with these chapters; and the greatness of the question at issue justifies even a heavier tax on the reader's attention. The matter is one which concerns each and all of us more than any other matter whatever. Though it affects us little in a direct way, the view we arrive at must indirectly affect us in all our relations—must determine our conception of the Universe, of Life, of Human Nature—must influence our ideas of right and wrong, and so modify our conduct. To reach that point of view from which the seeming discordance of Religion and Science disappears, and the two merge into one, must cause a revolution of thought fruitful in beneficial consequences, and must surely be worth an effort.

Here ending preliminaries, let us now address ourselves to this all-important inquiry.

CHAPTER II
ULTIMATE RELIGIOUS IDEAS

§ 9. When, on the sea-shore, we note how the hulls of distant vessels are hidden below the horizon, and how, of still remoter vessels, only the uppermost sails are visible, we realize with tolerable clearness the slight curvature of that portion of the sea's surface which lies before us. But when we seek in imagination to follow out this curved surface as it actually exists, slowly bending round until all its meridians meet in a point eight thousand miles below our feet, we find ourselves utterly baffled. We cannot conceive in its real form and magnitude even that small segment of our globe which extends a hundred miles on every side of us; much less the globe as a whole. The piece of rock on which we stand can be mentally represented with something like completeness: we find ourselves able to think of its top, its sides, and its under surface at the same time; or so nearly at the same time that they seem all present in consciousness together; and so we can form what we call a conception of the rock. But to do the like with the Earth we find impossible. If even to imagine the antipodes as at that distant place in space which it actually occupies, is beyond our power; much more beyond our power must it be at the same time to imagine all other remote points on the Earth's surface as in their actual places. Yet we habitually speak as though we had an idea of the Earth—as though we could think of it in the same way that we think of minor objects.

What conception, then, do we form of it? the reader may ask. That its name calls up in us some state of consciousness is unquestionable; and if this state of consciousness is not a conception, properly so called, what is it? The answer seems to be this:—We have learnt by indirect methods that the Earth is a sphere; we have formed models approximately representing its shape and the distribution of its parts; generally when the Earth is referred to, we either think of an indefinitely extended mass beneath our feet, or else, leaving out the actual Earth, we think of a body like a terrestrial globe; but when we seek to imagine the Earth as it really is, we join these two ideas as well as we can—such perception as our eyes give us of the Earth's surface we couple with the conception of a sphere. And thus we form of the Earth, not a conception properly so called, but only a symbolic conception.[6]

A large proportion of our conceptions, including all those of much generality, are of this order. Great magnitudes, great durations, great numbers, are none of them actually conceived, but are all of them conceived more or less symbolically; and so, too, are all those classes of objects of which we predicate some common fact. When mention is made of any individual man, a tolerably complete idea of him is formed. If the family he belongs to be spoken of, probably but a part of it will be represented in thought: under the necessity of attending to that which is said about the family, we realize in imagination only its most important or familiar members, and pass over the rest with a nascent consciousness which we know could, if requisite, be made complete. Should something be remarked of the class, say farmers, to which this family belongs, we neither enumerate in thought all the individuals contained in the class, nor believe that we could do so if required; but we are content with taking some few samples of it, and remembering that these could be indefinitely multiplied. Supposing the subject of which something is predicated be Englishmen, the answering state of consciousness is a still more inadequate representative of the reality. Yet more remote is the likeness of the thought to the thing, if reference be made to Europeans or to human beings. And when we come to propositions concerning the mammalia, or concerning the whole of the vertebrata, or concerning animals in general, or concerning all organic beings, the unlikeness of our conceptions to the objects named reaches its extreme. Throughout which series of instances we see, that as the number of objects grouped together in thought increases, the concept, formed of a few typical samples joined with the notion of multiplicity, becomes more and more a mere symbol; not only because it gradually ceases to represent the size of the group, but also because as the group grows more heterogeneous, the typical samples thought of are less like the average objects which the group contains.

This formation of symbolic conceptions, which inevitably arises as we pass from small and concrete objects to large and to discrete ones, is mostly a very useful, and indeed necessary, process. When, instead of things whose attributes can be tolerably well united in a single state of consciousness, we have to deal with things whose attributes are too vast or numerous to be so united, we must either drop in thought part of their attributes, or else not think of them at all—either form a more or less symbolic conception, or no conception. We must predicate nothing of objects too great or too multitudinous to be mentally represented; or we must make our predications by the help of extremely inadequate representations of such objects—mere symbols of them.

But while by this process alone we are enabled to form general propositions, and so to reach general conclusions, we are by this process perpetually led into danger, and very often into error. We habitually mistake our symbolic conceptions for real ones; and so are betrayed into countless false inferences. Not only is it that in proportion as the concept we form of any thing or class of things, misrepresents the reality, we are apt to be wrong in any assertion we make respecting the reality; but it is that we are led to suppose we have truly conceived a great variety of things which we have conceived only in this fictitious way; and further to confound with these certain things which cannot be conceived in any way. How almost unavoidably we fall into this error it will be needful here to observe.

From objects readily representable in their totality, to those of which we cannot form even an approximate representation, there is an insensible transition. Between a pebble and the entire Earth a series of magnitudes might be introduced, each of which differed from the adjacent ones so slightly that it would be impossible to say at what point in the series our conceptions of them became inadequate. Similarly, there is a gradual progression from those groups of a few individuals which we can think of as groups with tolerable completeness, to those larger and larger groups of which we can form nothing like true ideas. Whence it is manifest that we pass from actual conceptions to symbolic ones by infinitesimal steps. Note next that we are led to deal with our symbolic conceptions as though they were actual ones, not only because we cannot clearly separate the two, but also because, in the great majority of cases, the first serve our purposes nearly or quite as well as the last—are simply the abbreviated signs we substitute for those more elaborate signs which are our equivalents for real objects. Those very imperfect representations of ordinary things which we habitually make in thinking, we know can be developed into adequate ones if needful. Those concepts of larger magnitudes and more extensive classes which we cannot make adequate, we still find can be verified by some indirect process of measurement or enumeration. And even in the case of such an utterly inconceivable object as the Solar System, we yet, through the fulfilment of predictions founded on our symbolic conception of it, gain the conviction that this symbolic conception stands for an actual existence, and, in a sense, truly expresses certain of its constituent relations. Thus our symbolic conceptions being in the majority of cases capable of development into complete ones, and in most other cases serving as steps to conclusions which are proved valid by their correspondence with observation, we acquire a confirmed habit of dealing with them as true conceptions—as real representations of actualities. Learning by long experience that they can, if needful, be verified, we are led habitually to accept them without verification. And thus we open the door to some which profess to stand for known things, but which really stand for things that cannot be known in any way.

To sum up, we must say of conceptions in general, that they are complete only when the attributes of the object conceived are of such number and kind that they can be represented in consciousness so nearly at the same time as to seem all present together; that as the objects conceived become larger and more complex, some of the attributes first thought of fade from consciousness before the rest have been represented, and the conception thus becomes imperfect; that when the size, complexity, or discreteness of the object conceived becomes very great, only a small portion of its attributes can be thought of at once, and the conception formed of it thus becomes so inadequate as to be a mere symbol; that nevertheless such symbolic conceptions, which are indispensable in general thinking, are legitimate, provided that by some cumulative or indirect process of thought, or by the fulfilment of predictions based on them, we can assure ourselves that they stand for actualities; but that when our symbolic conceptions are such that no cumulative or indirect processes of thought can enable us to ascertain that there are corresponding actualities, nor any predictions be made whose fulfilment can prove this, then they are altogether vicious and illusive, and in no way distinguishable from pure fictions.

§ 10. And now to consider the bearings of this general truth on our immediate topic—Ultimate Religious Ideas.

To the aboriginal man and to every civilized child the problem of the Universe suggests itself. What is it? and whence comes it? are questions that press for solution, when, from time to time, the imagination rises above daily trivialities. To fill the vacuum of thought, any theory that is proposed seems better than none. And in the absence of others, any theory that is proposed easily gains a footing and afterwards maintains its ground: partly from the readiness of mankind to accept proximate explanations; partly from the authority which soon accumulates round such explanations when given.

A critical examination, however, will prove not only that no current hypothesis is tenable, but also that no tenable hypothesis can be framed.

§ 11. Respecting the origin of the Universe three verbally intelligible suppositions may be made. We may assert that it is self-existent; or that it is self-created; or that it is created by an external agency. Which of these suppositions is most credible it is not needful here to inquire. The deeper question, into which this finally merges, is, whether any one of them is even conceivable in the true sense of the word. Let us successively test them.

When we speak of a man as self-supporting, of an apparatus as self-acting, or of a tree as self-developed, our expressions, however inexact, stand for things that can be realized in thought with tolerable completeness. Our conception of the self-development of a tree is doubtless symbolic. But though we cannot really represent in consciousness the entire series of complex changes through which the tree passes, yet we can thus represent the leading features of the series; and general experience teaches us that by long continued observation we could gain the power to realize in thought a series of changes more fully representing the actual series: that is, we know that our symbolic conception of self-development can be expanded into something like a real conception; and that it expresses, however inaccurately, an actual process in nature. But when we speak of self-existence, and, helped by the above analogies, form some vague symbolic conception of it, we delude ourselves in supposing that this symbolic conception is of the same order as the others. On joining the word *self* to the word *existence*, the force of association makes us believe we have a thought like that suggested by the compound word self-acting. An endeavour to expand this symbolic conception, however, will undeceive us. In the first place, it is clear that by self-existence we especially mean, an existence independent of any other—not produced by any other: the assertion of self-existence is simply an indirect denial of creation. In thus excluding the idea of any antecedent cause, we necessarily exclude the idea of a beginning; for to admit the idea of a beginning—to admit that there was a time when the existence had not commenced—is to admit that its commencement was determined by something, or was caused; which is a contradiction. Self-existence, therefore, necessarily means existence without a beginning; and to form a conception of self-existence is to form a conception of existence without a beginning. Now by no mental effort can we do this. To conceive existence through infinite past-time, implies the conception of infinite past-time, which is an impossibility. To this let us add, that even were self-existence conceivable, it would not in any sense be an explanation of the Universe. No one will say that the existence of an object at the present moment is made easier to understand by the discovery that it existed an hour ago, or a day ago, or a year ago; and if its existence now is not made in the least degree more comprehensible by its existence during some previous finite period of time, then no accumulation of such finite periods, even could we extend them to an infinite period, would make it more comprehensible. Thus the Atheistic theory is not only absolutely unthinkable, but, even if it were thinkable, would not be a solution. The assertion that the Universe is self-existent does not really carry us a step beyond the cognition of its present existence; and so leaves us with a mere re-statement of the mystery.

The hypothesis of self-creation, which practically amounts to what is called Pantheism, is similarly incapable of being represented in thought. Certain phenomena, such as the precipitation of invisible vapour into cloud, aid us in forming a symbolic conception of a self-evolved Universe; and there are not wanting indications in the heavens, and on the earth, which help us to render this conception tolerably definite. But while the succession of phases through which the Universe has passed in reaching its present form, may perhaps be comprehended as in a sense self-determined; yet the impossibility of expanding our symbolic conception of self-creation into a real conception, remains as complete as ever. Really to conceive self-creation, is to conceive potential existence passing into actual existence by some inherent necessity; which we cannot do. We cannot form any idea of a potential existence of the universe, as distinguished from its actual existence. If represented in thought at all, potential existence must be represented as *something*, that is as an actual existence; to suppose that it can be represented as nothing, involves two absurdities—that nothing is more than a negation, and can be positively represented in thought; and that one nothing is distinguished from all other nothings by its power to develope into something. Nor is this all. We have no state of consciousness answering to the words—an inherent necessity by which potential existence became actual existence. To render them into thought, existence, having for an indefinite period remained in one form, must be conceived as passing without any external or additional impulse, into another form; and this involves the idea of a change without a cause—a thing of which no idea is possible. Thus the terms of this hypothesis do not stand for real thoughts; but merely suggest the vaguest symbols incapable of any interpretation. Moreover, even were it true that potential existence is conceivable as a different thing from actual existence; and that the transition from the one to the other can be mentally realized as a self-determined change; we should still be no forwarder: the problem would simply be removed a step back. For whence the potential existence? This would just as much require accounting for as actual existence; and just the same difficulties would meet us. Respecting the origin of such a latent power, no other suppositions could be made than those above named—self-existence, self-creation, creation by external agency. The self-existence of a potential universe is no more conceivable than we have found the self-existence of the actual universe to be. The self-creation of such a potential universe would involve over again the difficulties here stated—would imply behind this potential universe a more remote potentiality; and so on in an infinite series, leaving us at last no forwarder than at first. While to assign as the source of this potential universe an external agency, would be to introduce the notion of a potential universe for no purpose whatever.

There remains to be examined the commonly-received or theistic hypothesis—creation by external agency. Alike in the rudest creeds and in the cosmogony long current among ourselves, it is assumed that the genesis of the Heavens and the Earth is effected somewhat after the manner in which a workman shapes a piece of furniture. And this assumption is made not by theologians only, but by the immense majority of philosophers, past and present. Equally in the writings of Plato, and in those of not a few living men of science, we find it taken for granted that there is an analogy between the process of creation and the process of manufacture. Now in the first place, not only is this conception one that cannot by any cumulative process of thought, or the fulfilment of predictions based on it, be shown to answer to anything actual; and not only is it that in the absence of all evidence respecting the process of creation, we have no proof of correspondence even between this limited conception and some limited portion of the fact; but it is that the conception is not even consistent with itself—cannot be realized in thought, when all its assumptions are granted. Though it is true that the proceedings of a human artificer may vaguely symbolize to us a method after which the Universe might be shaped, yet they do not help us to comprehend the real mystery; namely, the origin of the material of which the Universe consists. The artizan does not make the iron, wood, or stone, he uses; but merely fashions and combines them. If we suppose suns, and planets, and satellites, and all they contain to have been similarly formed by a "Great Artificer," we suppose merely that certain pre-existing elements were thus put into their present arrangement. But whence the pre-existing elements? The comparison helps us not in the least to understand that; and unless it helps us to understand that, it is worthless. The production of matter out of nothing is the real mystery, which neither this simile nor any other enables us to conceive; and a simile which does not enable us to conceive this, may just as well be dispensed with. Still more manifest does the insufficiency of this theory of creation become, when we turn from material objects to that which contains them—when instead of matter we contemplate space. Did there exist nothing but an immeasurable void, explanation would be needed as much as now. There would still arise the question—how came it so? If the theory of creation by external agency were an adequate one, it would supply an answer; and its answer would be—space was made in the same manner that matter was made. But the impossibility of conceiving this is so manifest, that no one dares to assert it. For if space was created, it must have been previously non-existent. The non-existence of space cannot, however, by any mental effort be imagined. It is one of the most familiar truths that the idea of space as surrounding us on all sides, is not for a moment to be got rid of—not only are we compelled to think of space as now everywhere present, but we are unable to conceive its absence either in the past or the future. And if the non-existence of space is absolutely inconceivable, then, necessarily, its creation is absolutely inconceivable. Lastly, even supposing that the genesis of the Universe could really be represented in thought as the result of an external agency, the mystery would be as great as ever; for there would still arise the question—how came there to be an external agency? To account for this only the same three hypotheses are possible—self-existence, self-creation, and creation by external agency. Of these the last is useless: it commits us to an infinite series of such agencies, and even then leaves us where we were. By the second we are practically involved in the same predicament; since, as already shown, self-creation implies an infinite series of potential existences. We are obliged therefore to fall back upon the first, which is the one commonly accepted and commonly supposed to be satisfactory. Those who cannot conceive a self-existent universe; and who therefore assume a creator as the source of the universe; take for granted that they can conceive a self-existent creator. The mystery which they recognize in this great fact surrounding them on every side, they transfer to an alleged source of this great fact; and then suppose that they have solved the mystery. But they delude themselves. As was proved at the outset of the argument, self-existence is rigorously inconceivable; and this holds true whatever be the nature of the object of which it is predicated. Whoever agrees that the atheistic hypothesis is untenable because it involves the impossible idea of self-existence, must perforce admit that the theistic hypothesis is untenable if it contains the same impossible idea.

Thus these three different suppositions respecting the origin of things, verbally intelligible though they are, and severally seeming to their respective adherents quite rational, turn out, when critically examined, to be literally unthinkable. It is not a question of probability, or credibility, but of conceivability. Experiment proves that the elements of these hypotheses cannot even be put together in consciousness; and we can entertain them only as we entertain such pseud-ideas as a square fluid and a moral substance— only by abstaining from the endeavour to render them into actual thoughts. Or, reverting to our original mode of statement, we may say that they severally involve symbolic conceptions of the illegitimate and illusive kind. Differing so widely as they seem to do, the atheistic, the pantheistic, and the theistic hypotheses contain the same ultimate element. It is impossible to avoid making the assumption of self-existence somewhere; and whether that assumption be made nakedly, or under complicated disguises, it is equally vicious, equally unthinkable. Be it a fragment of matter, or some fancied potential form of matter, or some more remote and still less imaginable cause, our conception of its self-existence can be formed only by joining with it the notion of unlimited duration through past time. And as unlimited duration is inconceivable, all those formal ideas into which it enters are inconceivable; and indeed, if such an expression is allowable, are the more inconceivable in proportion as the other elements of the ideas are indefinite. So that in fact, impossible as it is to think of the actual universe as self-existing, we do but multiply impossibilities of thought by every attempt we make to explain its existence.

§ 12. If from the origin of the Universe we turn to its nature, the like insurmountable difficulties rise up before us on all sides—or rather, the same difficulties under new aspects. We find ourselves on the one hand obliged to make certain assumptions; and yet on the other hand we find these assumptions cannot be represented in thought.

When we inquire what is the meaning of the various effects produced upon our senses—when we ask how there come to be in our consciousness impressions of sounds, of colours, of tastes, and of those various attributes which we ascribe to bodies; we are compelled to regard them as the effects of some cause. We may stop short in the belief that this cause is what we call matter. Or we may conclude, as some do, that matter is only a certain mode of manifestation of spirit; which is therefore the true cause. Or, regarding matter and spirit as proximate agencies, we may attribute all the changes wrought in our consciousness to immediate divine power. But be the cause we assign what it may, we are obliged to suppose *some* cause. And we are not only obliged to suppose some cause, but also a first cause. The matter, or spirit, or whatever we assume to be the agent producing on us these various impressions, must either be the first cause of them or not. If it is the first cause, the conclusion is reached. If it is not the first cause, then by implication there must be a cause behind it; which thus becomes the real cause of the effect. Manifestly, however complicated the assumptions, the same conclusion must inevitably be reached. We cannot think at all about the impressions which the external world produces on us, without thinking of them as caused; and we cannot carry out an inquiry concerning their causation, without inevitably committing ourselves to the hypothesis of a First Cause.

But now if we go a step further, and ask what is the nature of this First Cause, we are driven by an inexorable logic to certain further conclusions. Is the First Cause finite or infinite? If we say finite we involve ourselves in a dilemma. To think of the First Cause as finite, is to think of it as limited. To think of it as limited, necessarily implies a conception of something beyond its limits: it is absolutely impossible to conceive a thing as bounded without conceiving a region surrounding its boundaries. What now must we say of this region? If the First Cause is limited, and there consequently lies something outside of it, this something must have no First Cause—must be uncaused. But if we admit that there can be something uncaused, there is no reason to assume a cause for anything. If beyond that finite region over which the First Cause extends, there lies a region, which we are compelled to regard as infinite, over which it does not extend—if we admit that there is an infinite uncaused surrounding the finite caused; we tacitly abandon the hypothesis of causation altogether. Thus it is impossible to consider the First Cause as finite. And if it cannot be finite it must be infinite.

Another inference concerning the First Cause is equally unavoidable. It must be independent. If it is dependent it cannot be the First Cause; for that must be the First Cause on which it depends. It is not enough to say that it is partially independent; since this implies some necessity which determines its partial dependence, and this necessity, be it what it may, must be a higher cause, or the true First Cause, which is a contradiction. But to think of the First Cause as totally independent, is to think of it as that which exists in the absence of all other existence; seeing that if the presence of any other existence is necessary, it must be partially dependent on that other existence, and so cannot be the First Cause. Not only however must the First Cause be a form of being which has no necessary relation to any other form of being, but it can have no necessary relation within itself. There can be nothing in it which determines change, and yet nothing which prevents change. For if it contains something which imposes such necessities or restraints, this something must be a cause higher than the First Cause, which is absurd. Thus the First Cause must be in every sense perfect, complete, total: including within itself all power, and transcending all law. Or to use the established word, it must be absolute.

Here then respecting the nature of the Universe, we seem committed to certain unavoidable conclusions. The objects and actions surrounding us, not less than the phenomena of our own consciousness, compel us to ask a cause; in our search for a cause, we discover no resting place until we arrive at the hypothesis of a First Cause; and we have no alternative but to regard this First Cause as Infinite and Absolute. These are inferences forced upon us by arguments from which there appears no escape. It is hardly needful however to show those who have followed thus far, how illusive are these reasonings and their results. But that it would tax the reader's patience to no purpose, it might easily be proved that the materials of which the argument is built, equally with the conclusions based on them, are merely symbolic conceptions of the illegitimate order. Instead, however, of repeating the disproof used above, it will be desirable to pursue another method; showing the fallacy of these conclusions by disclosing their mutual contradictions.

Here I cannot do better than avail myself of the demonstration which Mr Mansel, carrying out in detail the doctrine of Sir William Hamilton, has given in his "Limits of Religious Thought." And I gladly do this, not only because his mode of presentation cannot be improved, but also because, writing as he does in defence of the current Theology, his reasonings will be the more acceptable to the majority of readers.

§ 13. Having given preliminary definitions of the First Cause, of the Infinite, and of the Absolute, Mr Mansel says:—

"But these three conceptions, the Cause, the Absolute, the Infinite, all equally indispensable, do they not imply contradiction to each other, when viewed in conjunction, as attributes of one and the same Being? A Cause cannot, as such, be absolute: the Absolute cannot, as such, be a cause. The cause, as such, exists only in relation to its effect: the cause is a cause of the effect; the effect is an effect of the cause. On the other hand, the conception of the Absolute implies a possible existence out of all relation. We attempt to escape from this apparent contradiction, by introducing the idea of succession in time. The Absolute exists first by itself, and afterwards becomes a Cause. But here we are checked by the third conception, that of the Infinite. How can the Infinite become that which it was not from the first? If Causation is a possible mode of existence, that which exists without causing is not infinite; that which becomes a cause has passed beyond its former limits." * * *

"Supposing the Absolute to become a cause, it will follow that it operates by means of freewill and consciousness. For a necessary cause cannot be conceived as absolute and infinite. If necessitated by something beyond itself, it is thereby limited by a superior power; and if necessitated by itself, it has in its own nature a necessary relation to its effect. The act of causation must therefore be voluntary; and volition is only possible in a conscious being. But consciousness again is only conceivable as a relation. There must be a conscious subject, and an object of which he is conscious. The subject is a subject to the object; the object is an object to the subject; and neither can exist by itself as the absolute. This difficulty, again, may be for the moment evaded, by distinguishing between the absolute as related to another and the absolute as related to itself. The Absolute, it may be said, may possibly be conscious, provided it is only conscious of itself. But this alternative is, in ultimate analysis, no less self-destructive than the other. For the object of consciousness, whether a mode of the subject's existence or not, is either created in and by the act of consciousness, or has an existence independent of it. In the former case, the object depends upon the subject, and the subject alone is the true absolute. In the latter case, the subject depends upon the object, and the object alone is the true absolute. Or if we attempt a third hypothesis, and maintain that each exists independently of the other, we have no absolute at all, but only a pair of relatives; for coexistence, whether in consciousness or not, is itself a relation."

"The corollary from this reasoning is obvious. Not only is the Absolute, as conceived, incapable of a necessary relation to anything else; but it is also incapable of containing, by the constitution of its own nature, an essential relation within itself; as a whole, for instance, composed of parts, or as a substance consisting of attributes, or as a conscious subject in antithesis to an object. For if there is in the absolute any principle of unity, distinct from the mere accumulation of parts or attributes, this principle alone is the true absolute. If, on the other hand, there is no such principle, then there is no absolute at all, but only a plurality of relatives. The almost unanimous voice of philosophy, in pronouncing that the absolute is both one and simple, must be accepted as the voice of reason also, so far as reason has any voice in the matter. But this absolute unity, as indifferent and containing no attributes, can neither be distinguished from the multiplicity of finite beings by any characteristic feature, nor be identified with them in their multiplicity. Thus we are landed in an inextricable dilemma. The Absolute cannot be conceived as conscious, neither can it be conceived as unconscious: it cannot be conceived as complex, neither can it be conceived as simple: it cannot be conceived by difference, neither can it be conceived by the absence of difference: it cannot be identified with the universe, neither can it be distinguished from it. The One and the Many, regarded as the beginning of existence, are thus alike incomprehensible."

"The fundamental conceptions of Rational Theology being thus self-destructive, we may naturally expect to find the same antagonism manifested in their special applications. * * * How, for example, can Infinite Power be able to do all things, and yet Infinite Goodness be unable to do evil? How can Infinite Justice exact the utmost penalty for every sin, and yet Infinite Mercy pardon the sinner? How can Infinite Wisdom know all that is to come, and yet Infinite Freedom be at liberty to do or to forbear? How is the existence of Evil compatible with that of an infinitely perfect Being; for if he wills it, he is not infinitely good; and if he wills it not, his will is thwarted and his sphere of action limited?" * * *

"Let us, however, suppose for an instant that these difficulties are surmounted, and the existence of the Absolute securely established on the testimony of reason. Still we have not succeeded in reconciling this idea with that of a Cause: we have done nothing towards explaining how the absolute can give rise to the relative, the infinite to the finite. If the condition of casual activity is a higher state than that of quiescence, the Absolute, whether acting voluntarily or involuntarily, has passed from a condition of comparative imperfection to one of comparative perfection; and therefore was not originally perfect. If the state of activity is an inferior state to that of quiescence, the Absolute, in becoming a cause, has lost its original perfection. There remains only the supposition that the two states are equal, and the act of creation one of complete indifference. But this supposition annihilates the unity of the absolute, or it annihilates itself. If the act of creation is real, and yet indifferent, we must admit the possibility of two conceptions of the absolute, the one as productive, the other as non-productive. If the act is not real, the supposition itself vanishes."

"Again, how can the relative be conceived as coming into being? If it is a distinct reality from the absolute, it must be conceived as passing from non-existence into existence. But to conceive an object as non-existent, is again a self-contradiction; for that which is conceived exists, as an object of thought, in and by that conception. We may abstain from thinking of an object at all; but, if we think of it, we cannot but think of it as existing. It is possible at one time not to think of an object at all, and at another to think of it as already in being; but to think of it in the act of becoming, in the progress from not being into being, is to think that which, in the very thought, annihilates itself."

"To sum up briefly this portion of my argument. The conception of the Absolute and Infinite, from whatever side we view it, appears encompassed with contradictions. There is a contradiction in supposing such an object to exist, whether alone or in conjunction with others; and there is a contradiction in supposing it not to exist. There is a contradiction in conceiving it as one; and there is a contradiction in conceiving it as many. There is a contradiction in conceiving it as personal; and there is a contradiction in conceiving it as impersonal. It cannot, without contradiction, be represented as active; nor, without equal contradiction, be represented as inactive. It cannot be conceived as the sum of all existence; nor yet can it be conceived as a part only of that sum."

§ 14. And now what is the bearing of these results on the question before us? Our examination of Ultimate Religious Ideas has been carried on with the view of making manifest some fundamental verity contained in them. Thus far however we have arrived at negative conclusions only. Criticising the essential conceptions involved in the different orders of beliefs, we find no one of them to be logically defensible. Passing over the consideration of credibility, and confining ourselves to that of conceivability, we see that Atheism, Pantheism, and Theism, when rigorously analysed, severally prove to be absolutely unthinkable. Instead of disclosing a fundamental verity existing in each, our investigation seems rather to have shown that there is no fundamental verity contained in any. To carry away this conclusion, however, would be a fatal error; as we shall shortly see.

Leaving out the accompanying moral code, which is in all cases a supplementary growth, every Religion may be defined as an *à priori* theory of the Universe. The surrounding facts being given, some form of agency is alleged which, in the opinion of those alleging it, accounts for these facts. Be it in the rudest Fetishism, which assumes a separate personality behind every phenomenon; be it in Polytheism, in which these personalities are partially generalized; be it in Monotheism, in which they are wholly generalized; or be it in Pantheism, in which the generalized personality becomes one with the phenomena; we equally find an hypothesis which is supposed to render the Universe comprehensible. Nay, even that which is commonly regarded as the negation of all Religion—even positive Atheism, comes within the definition; for it, too, in asserting the self-existence of Space, Matter, and Motion, which it regards as adequate causes of every appearance, propounds an *à priori* theory from which it holds the facts to be deducible. Now every theory tacitly asserts two things: firstly, that there is something to be explained; secondly, that such and such is the explanation. Hence, however widely different speculators may disagree in the solutions they give of the same problem; yet by implication they agree that there is a problem to be solved. Here then is an element which all creeds have in common. Religions diametrically opposed in their overt dogmas, are yet perfectly at one in the tacit conviction that the existence of the world with all it contains and all which surrounds it, is a mystery ever pressing for interpretation. On this point, if on no other, there is entire unanimity.

Thus we come within sight of that which we seek. In the last chapter, reasons were given for inferring that human beliefs in general, and especially the perennial ones, contain, under whatever disguises of error, some soul of truth; and here we have arrived at a truth underlying even the grossest superstitions. We saw further that this soul of truth was most likely to be some constituent common to conflicting opinions of the same order; and here we have a constituent which may be claimed alike by all religions. It was pointed out that this soul of truth would almost certainly be more abstract than any of the beliefs involving it; and the truth we have arrived at is one exceeding in abstractness the most abstract religious doctrines. In every respect, therefore, our conclusion answers to the requirements. It has all the characteristics which we inferred must belong to that fundamental verity expressed by religions in general.

That this is the vital element in all religions is further proved by the fact, that it is the element which not only survives every change, but grows more distinct the more highly the religion is developed. Aboriginal creeds, though pervaded by the idea of personal agencies which are usually unseen, yet conceive these agencies under perfectly concrete and ordinary forms— class them with the visible agencies of men and animals; and so hide a vague perception of mystery in disguises as unmysterious as possible. The Polytheistic conceptions in their advanced phases, represent the presiding personalities in greatly idealized shapes, existing in a remote region, working in subtle ways, and communicating with men by omens or through inspired persons; that is, the ultimate causes of things are regarded as less familiar and comprehensible. The growth of a Monotheistic faith, accompanied as it is by a denial of those beliefs in which the divine nature is assimilated to the human in all its lower propensities, shows us a further step in the same direction; and however imperfectly this higher faith is at first realized, we yet see in altars "to the unknown and unknowable God," and in the worship of a God that cannot by any searching be found out, that there is a clearer recognition of the inscrutableness of creation. Further developments of theology, ending in such assertions as that "a God understood would be no God at all," and "to think that God is, as we can think him to be, is blasphemy," exhibit this recognition still more distinctly; and it pervades all the cultivated theology of the present day. Thus while other constituents of religious creeds one by one drop away, this remains and grows even more manifest; and so is shown to be the essential constituent.

Nor does the evidence end here. Not only is the omnipresence of something which passes comprehension, that most abstract belief which is common to all religions, which becomes the more distinct in proportion as they develope, and which remains after their discordant elements have been mutually cancelled; but it is that belief which the most unsparing criticism of each leaves unquestionable—or rather makes ever clearer. It has nothing to fear from the most inexorable logic; but on the contrary is a belief which the most inexorable logic shows to be more profoundly true than any religion supposes. For every religion, setting out though it does with the tacit assertion of a mystery, forthwith proceeds to give some solution of this mystery; and so asserts that it is not a mystery passing human comprehension. But an examination of the solutions they severally propound, shows them to be uniformly invalid. The analysis of every possible hypothesis proves, not simply that no hypothesis is sufficient, but that no hypothesis is even thinkable. And thus the mystery which all religions recognize, turns out to be a far more transcendent mystery than any of them suspect—not a relative, but an absolute mystery.

Here, then, is an ultimate religious truth of the highest possible certainty—a truth in which religions in general are at one with each other, and with a philosophy antagonistic to their special dogmas. And this truth, respecting which there is a latent agreement among all mankind from the fetish-worshipper to the most stoical critic of human creeds, must be the one we seek. If Religion and Science are to be reconciled, the basis of reconciliation must be this deepest, widest, and most certain of all facts—that the Power which the Universe manifests to us is utterly inscrutable.

> 6. Those who may have before met with this term, will perceive that it is here used in quite a different sense.

CHAPTER III
ULTIMATE SCIENTIFIC IDEAS

§ 15. What are Space and Time? Two hypotheses are current respecting them: the one that they are objective, and the other that they are subjective — the one that they are external to, and independent of, ourselves, the other that they are internal, and appertain to our own consciousness. Let us see what becomes of these hypotheses under analysis.

To say that Space and Time exist objectively, is to say that they are entities. The assertion that they are non-entities is self-destructive: non-entities are non-existences; and to allege that non-existences exist objectively, is a contradiction in terms. Moreover, to deny that Space and Time are things, and so by implication to call them nothings, involves the absurdity that there are two kinds of nothing. Neither can they be regarded as attributes of some entity; seeing, not only that it is impossible really to conceive any entity of which they are attributes, but seeing further that we cannot think of them as disappearing, even if everything else disappeared; whereas attributes necessarily disappear along with the entities they belong to. Thus as Space and Time cannot be either non-entities, nor the attributes of entities, we have no choice but consider them as entities. But while, on the hypothesis of their objectivity, Space and Time must be classed as things, we find, on experiment, that to represent them in thought as things is impossible. To be conceived at all, a thing must be conceived as having attributes. We can distinguish something from nothing, only by the power which the something has to act on our consciousness; the several affections it produces on our consciousness (or else the hypothetical causes of them), we attribute to it, and call its attributes; and the absence of these attributes is the absence of the terms in which the something is conceived, and involves the absence of a conception. What now are the attributes of Space? The only one which it is possible for a moment to think of as belonging to it, is that of extension; and to credit it with this implies a confusion of thought. For extension and Space are convertible terms: by extension, as we ascribe it to surrounding objects, we mean occupancy of Space; and thus to say that Space is extended, is to say that Space occupies Space. How we are similarly unable to assign any attribute to Time, scarcely needs pointing out. Nor are Time and Space unthinkable as entities only from the absence of attributes; there is another peculiarity, familiar to readers of metaphysics, which equally excludes them from the category. All entities which we actually know as such, are limited; and even if we suppose ourselves either to know or to be able to conceive some unlimited entity, we of necessity in so classing it positively separate it from the class of limited entities. But of Space and Time we cannot assert either limitation or the absence of limitation. We find ourselves totally unable to form any mental image of unbounded Space; and yet totally unable to imagine bounds beyond which there is no Space. Similarly at the other extreme: it is impossible to think of a limit to the divisibility of Space; yet equally impossible to think of its infinite divisibility. And, without stating them, it will be seen that we labour under like impotencies in respect to Time. Thus we cannot conceive Space and Time as entities, and are equally disabled from conceiving them as either the attributes of entities or as non-entities. We are compelled to think of them as existing; and yet cannot bring them within those conditions under which existences are represented in thought.

Shall we then take refuge in the Kantian doctrine? shall we say that Space and Time are forms of the intellect, — "*à priori* laws or conditions of the conscious mind"? To do this is to escape from great difficulties by rushing into greater. The proposition with which Kant's philosophy sets out, verbally intelligible though it is, cannot by any effort be rendered into thought — cannot be interpreted into an idea properly so called, but stands merely for a pseud-idea. In the first place, to assert that Space and Time, as we are conscious of them, are subjective conditions, is by implication to assert that they are not objective realities: if the Space and Time present to our minds belong to the *ego*, then of necessity they do not belong to the *non-ego*. Now it is absolutely impossible to think this. The very fact on which Kant bases his hypothesis — namely that our consciousness of Space and Time cannot be suppressed — testifies as much; for that consciousness of Space and Time which we cannot rid ourselves of, is the consciousness of them as existing objectively. It is useless to reply that such an inability must inevitably result if they are subjective forms. The question here is — What does consciousness directly testify? And the direct testimony of consciousness is, that Time and Space are not within but without the mind; and so absolutely independent of it that they cannot be conceived to become non-existent even were the mind to become non-existent. Besides being positively unthinkable in what it tacitly denies, the theory of Kant is equally unthinkable in what it openly affirms. It is not simply that we cannot combine the thought of Space with the thought of our own personality, and contemplate the one as a property of the other — though our inability to do this would prove the inconceivableness of the hypothesis—but it is that the hypothesis carries in itself the proof of its own inconceivableness. For if Space and Time are forms of thought, they can never be thought of; since it is impossible for anything to be at once the *form* of thought and the *matter* of thought. That Space and Time are objects of consciousness, Kant emphatically asserts by saying that it is impossible to suppress the consciousness of them. How then, if they are *objects* of consciousness, can they at the same time be *conditions* of consciousness? If Space and Time are the conditions under which we think, then when we think of Space and Time themselves, our thoughts must be unconditioned; and if there can thus be unconditioned thoughts, what becomes of the theory?

It results therefore that Space and Time are wholly incomprehensible. The immediate knowledge which we seem to have of them, proves, when examined, to be total ignorance. While our belief in their objective reality is insurmountable, we are unable to give any rational account of it. And to posit the alternative belief (possible to state but impossible to realize) is merely to multiply irrationalities.

§ 16. Were it not for the necessities of the argument, it would be inexcusable to occupy the reader's attention with the threadbare, and yet unended, controversy respecting the divisibility of matter. Matter is either infinitely divisible or it is not: no third possibility can be named. Which of the alternatives shall we accept? If we say that Matter is infinitely divisible, we commit ourselves to a supposition not realizable in thought. We can bisect and re-bisect a body, and continually repeating the act until we reduce its parts to a size no longer physically divisible, may then mentally continue the process without limit. To do this, however, is not really to conceive the infinite divisibility of matter, but to form a symbolic conception incapable of expansion into a real one, and not admitting of other verification. Really to conceive the infinite divisibility of matter, is mentally to follow out the divisions to infinity; and to do this would require infinite time. On the other hand, to assert that matter is not infinitely divisible, is to assert that it is reducible to parts which no conceivable power can divide; and this verbal supposition can no more be represented in thought than the other. For each of such ultimate parts, did they exist, must have an under and an upper surface, a right and a left side, like any larger fragment. Now it is impossible to imagine its sides so near that no plane of section can be conceived between them; and however great be the assumed force of cohesion, it is impossible to shut out the idea of a greater force capable of overcoming it. So that to human intelligence the one hypothesis is no more acceptable than the other; and yet the conclusion that one or other must agree with the fact, seems to human intelligence unavoidable.

Again, leaving this insoluble question, let us ask whether substance has, in reality, anything like that extended solidity which it presents to our consciousness. The portion of space occupied by a piece of metal, seems to eyes and fingers perfectly filled: we perceive a homogeneous, resisting mass, without any breach of continuity. Shall we then say that Matter is as actually solid as it appears? Shall we say that whether it consists of an infinitely divisible element or of ultimate units incapable of further division, its parts are everywhere in actual contact? To assert as much entangles us in insuperable difficulties. Were Matter thus absolutely solid, it would be, what it is not—absolutely incompressible; since compressibility, implying the nearer approach of constituent parts, is not thinkable unless there is unoccupied space between the parts. Nor is this all. It is an established mechanical truth, that if a body, moving at a given velocity, strikes an equal body at rest in such wise that the two move on together, their joint velocity will be but half that of the striking body. Now it is a law of which the negation is inconceivable, that in passing from any one degree of magnitude to any other, all intermediate degrees must be passed through. Or, in the case before us, a body moving at velocity 4, cannot, by collision, be reduced to velocity 2, without passing through all velocities between 4 and 2. But were Matter truly solid—were its units absolutely incompressible and in absolute contact—this "law of continuity," as it is called, would be broken in every case of collision. For when, of two such units, one moving at velocity 4 strikes another at rest, the striking unit must have its velocity 4 instantaneously reduced to velocity 2; must pass from velocity 4 to velocity 2 without any lapse of time, and without passing through intermediate velocities; must be moving with velocities 4 and 2 at the same instant, which is impossible.

The supposition that Matter is absolutely solid being untenable, there presents itself the Newtonian supposition, that it consists of solid atoms not in contact but acting on each other by attractive and repulsive forces, varying with the distances. To assume this, however, merely shifts the difficulty: the problem is simply transferred from the aggregated masses of matter to these hypothetical atoms. For granting that Matter, as we perceive it, is made up of such dense extended units surrounded by atmospheres of force, the question still arises—What is the constitution of these units? We have no alternative but to regard each of them as a small piece of matter. Looked at through a mental microscope, each becomes a mass of substance such as we have just been contemplating. Exactly the same inquiries may be made respecting the parts of which each atom consists; while exactly the same difficulties stand in the way of every answer. And manifestly, even were the hypothetical atom assumed to consist of still minuter ones, the difficulty would re-appear at the next step; nor could it be got rid of even by an infinite series of such assumptions.

Boscovich's conception yet remains to us. Seeing that Matter could not, as Leibnitz suggested, be composed of unextended monads (since the juxtaposition of an infinity of points having no extension, could not produce that extension which matter possesses); and perceiving objections to the view entertained by Newton; Boscovich proposed an intermediate theory, uniting, as he considered, the advantages of both and avoiding their difficulties. His theory is, that the constituents of Matter are centres of force—points without dimensions, which attract and repel each other in suchwise as to be kept at specific distances apart. And he argues, mathematically, that the forces possessed by such centres might so vary with the distances, that under given conditions the centres would remain in stable equilibrium with definite interspaces; and yet, under other conditions, would maintain larger or smaller interspaces. This speculation however, ingeniously as it is elaborated, and eluding though it does various difficulties, posits a proposition which cannot by any effort be represented in thought: it escapes all the inconceivabilities above indicated, by merging them in the one inconceivability with which it sets out. A centre of force absolutely without extension is unthinkable: answering to these words we can form nothing more than a symbolic conception of the illegitimate order. The idea of resistance cannot be separated in thought from the idea of an extended body which offers resistance. To suppose that central forces can reside in points not infinitesimally small but occupying no space whatever—points having position only, with nothing to mark their position—points in no respect distinguishable from the surrounding points that are not centres of force;—to suppose this, is utterly beyond human power.

Here it may possibly be said, that though all hypotheses respecting the constitution of Matter commit us to inconceivable conclusions when logically developed, yet we have reason to think that one of them corresponds with the fact. Though the conception of Matter as consisting of dense indivisible units, is symbolic and incapable of being completely thought out, it may yet be supposed to find indirect verification in the truths of chemistry. These, it is argued, necessitate the belief that Matter consists of particles of specific weights, and therefore of specific sizes. The general law of definite proportions seems impossible on any other condition than the existence of ultimate atoms; and though the combining weights of the respective elements are termed by chemists their "equivalents," for the purpose of avoiding a questionable assumption, we are unable to think of the combination of such definite weights, without supposing it to take place between definite numbers of definite particles. And thus it would appear that the Newtonian view is at any rate preferable to that of Boscovich. A disciple of Boscovich, however, may reply that his master's theory is involved in that of Newton; and cannot indeed be escaped. "What," he may ask, "is it that holds together the parts of these ultimate atoms?". "A cohesive force," his opponent must answer. "And what," he may continue, "is it that holds together the parts of any fragments into which, by sufficient force, an ultimate atom might be broken?" Again the answer must be—a cohesive force. "And what," he may still ask, "if the ultimate atom were, as we can imagine it to be, reduced to parts as small in proportion to it, as it is in proportion to a tangible mass of matter—what must give each part the ability to sustain itself, and to occupy space?" Still there is no answer but—a cohesive force. Carry the process in thought as far as we may, until the extension of the parts is less than can be imagined, we still cannot escape the admission of forces by which the extension is upheld; and we can find no limit until we arrive at the conception of centres of force without any extension.

Matter then, in its ultimate nature, is as absolutely incomprehensible as Space and Time. Frame what suppositions we may, we find on tracing out their implications that they leave us nothing but a choice between opposite absurdities.

§ 17. A body impelled by the hand is clearly perceived to move, and to move in a definite direction: there seems at first sight no possibility of doubting that its motion is real, or that it is towards a given point. Yet it is easy to show that we not only may be, but usually are, quite wrong in both these judgments. Here, for instance, is a ship which, for simplicity's sake, we will suppose to be anchored at the equator with her head to the West. When the captain walks from stem to stern, in what direction does he move? East is the obvious answer—an answer which for the moment may pass without criticism. But now the anchor is heaved, and the vessel sails to the West with a velocity equal to that at which the captain walks. In what direction does he now move when he goes from stem to stern? You cannot say East, for the vessel is carrying him as fast towards the West as he walks to the East; and you cannot say West for the converse reason. In respect to surrounding space he is stationary; though to all on board the ship he seems to be moving. But now are we quite sure of this conclusion?—Is he really stationary? When we take into account the Earth's motion round its axis, we find that instead of being stationary he is travelling at the rate of 1000 miles per hour to the East; so that neither the perception of one who looks at him, nor the inference of one who allows for the ship's motion, is anything like the truth. Nor indeed, on further consideration, shall we find this revised conclusion to be much better. For we have forgotten to allow for the Earth's motion in its orbit. This being some 68,000 miles per hour, it follows that, assuming the time to be midday, he is moving, not at the rate of 1000 miles per hour to the East, but at the rate of 67,000 miles per hour to the West. Nay, not even now have we discovered the true rate and the true direction of his movement. With the Earth's progress in its orbit, we have to join that of the whole Solar system towards the constellation Hercules; and when we do this, we perceive that he is moving neither East nor West, but in a line inclined to the plane of the Ecliptic, and at a velocity greater or less (according to the time of the year) than that above named. To which let us add, that were the dynamic arrangements of our sidereal system fully known to us, we should probably discover the direction and rate of his actual movement to differ considerably even from these. How illusive are our ideas of Motion, is thus made sufficiently manifest. That which seems moving proves to be stationary; that which seems stationary proves to be moving; while that which we conclude to be going rapidly in one direction, turns out to be going much more rapidly in the opposite direction. And so we are taught that what we are conscious of is not the real motion of any object, either in its rate or direction; but merely its motion as measured from an assigned position—either the position we ourselves occupy or some other. Yet in this very process of concluding that the motions we perceive are not the real motions, we tacitly assume that there are real motions. In revising our successive judgments concerning a body's course or velocity, we take for granted that there is an actual course and an actual velocity—we take for granted that there are fixed points in space with respect to which all motions are absolute; and we find it impossible to rid ourselves of this idea. Nevertheless, absolute motion cannot even be imagined, much less known. Motion as taking place apart from those limitations of space which we habitually associate with it, is totally unthinkable. For motion is change of place; but in unlimited space, change of place is inconceivable, because place itself is inconceivable. Place can be conceived only by reference to other places; and in the absence of objects dispersed through space, a place could be conceived only in relation to the limits of space; whence it follows that in unlimited space, place cannot be conceived—all places must be equidistant from boundaries that do not exist. Thus while we are obliged to think that there is an absolute motion, we find absolute motion incomprehensible.

Another insuperable difficulty presents itself when we contemplate the transfer of Motion. Habit blinds us to the marvelousness of this phenomenon. Familiar with the fact from childhood, we see nothing remarkable in the ability of a moving thing to generate movement in a thing that is stationary. It is, however, impossible to understand it. In what respect does a body after impact differ from itself before impact? What is this added to it which does not sensibly affect any of its properties and yet enables it to traverse space? Here is an object at rest and here is the same object moving. In the one state it has no tendency to change its place; but in the other it is obliged at each instant to assume a new position. What is it which will for ever go on producing this effect without being exhausted? and how does it dwell in the object? The motion you say has been communicated. But how?—What has been communicated? The striking body has not transferred a *thing* to the body struck; and it is equally out of the question to say that it has transferred an *attribute*. What then has it transferred?

Once more there is the old puzzle concerning the connexion between Motion and Rest. We daily witness the gradual retardation and final stoppage of things projected from the hand or otherwise impelled; and we equally often witness the change from Rest to Motion produced by the application of force. But truly to represent these transitions in thought, we find impossible. For a breach of the law of continuity seems necessarily involved; and yet no breach of it is conceivable. A body travelling at a given velocity cannot be brought to a state of rest, or no velocity, without passing through all intermediate velocities. At first sight nothing seems easier than to imagine it doing this. It is quite possible to think of its motion as diminishing insensibly until it becomes infinitesimal; and many will think equally possible to pass in thought from infinitesimal motion to no motion. But this is an error. Mentally follow out the decreasing velocity as long as you please, and there still remains *some* velocity. Halve and again halve the rate of movement for ever, yet movement still exists; and the smallest movement is separated by an impassable gap from no movement. As something, however minute, is infinitely great in comparison with nothing; so is even the least conceivable motion, infinite as compared with rest. The converse perplexities attendant on the transition from Rest to Motion, need not be specified. These, equally with the foregoing, show us that though we are obliged to think of such changes as actually occurring, their occurrence cannot be realized.

Thus neither when considered in connexion with Space, nor when considered in connexion with Matter, nor when considered in connexion with Rest, do we find that Motion is truly cognizable. All efforts to understand its essential nature do but bring us to alternative impossibilities of thought.

§ 18. On lifting a chair, the force exerted we regard as equal to that antagonistic force called the weight of the chair; and we cannot think of these as equal without thinking of them as like in kind; since equality is conceivable only between things that are connatural. The axiom that action and reaction are equal and in opposite directions, commonly exemplified by this very instance of muscular effort *versus* weight, cannot be mentally realized on any other condition. Yet, contrariwise, it is incredible that the force as existing in the chair really resembles the force as present to our minds. It scarcely needs to point out that the weight of the chair produces in us various feelings according as we support it by a single finger, or the whole hand, or the leg; and hence to argue that as it cannot be like all these sensations there is no reason to believe it like any. It suffices to remark that since the force as known to us is an affection of consciousness, we cannot conceive the force existing in the chair under the same form without endowing the chair with consciousness. So that it is absurd to think of Force as in itself like our sensation of it, and yet necessary so to think of it if we realize it in consciousness at all.

How, again, can we understand the connexion between Force and Matter? Matter is known to us only through its manifestations of Force: our ultimate test of Matter is the ability to resist: abstract its resistance and there remains nothing but empty extension. Yet, on the other hand, resistance is equally unthinkable apart from Matter—apart from something extended. Not only, as pointed out some pages back, are centres of force devoid of extension unimaginable; but, as an inevitable corollary, we cannot imagine either extended or unextended centres of force to attract and repel other such centres at a distance, without the intermediation of some kind of matter. We have here to remark, what could not without anticipation be remarked when treating of Matter, that the hypothesis of Newton, equally with that of Boscovich, is open to the charge that it supposes one thing to act upon another through a space which is absolutely empty—a supposition which cannot be represented in thought. This charge is indeed met by the introduction of a hypothetical fluid existing between the atoms or centres. But the problem is not thus solved: it is simply shifted, and re-appears when the constitution of this fluid is inquired into. How impossible it is to elude the difficulty presented by the transfer of Force through space, is best seen in the case of astronomical forces. The Sim acts upon us in such way as to produce the sensations of light and heat; and we have ascertained that between the cause as existing in the Sun, and the effect as experienced on the Earth, a lapse of about eight minutes occurs: whence unavoidably result in us, the conceptions of both a force and a motion. So that for the assumption of a luminiferous ether, there is the defence, not only that the exercise of force through 95,000,000 of miles of absolute vacuum is inconceivable, but also that it is impossible to conceive motion in the absence of something moved. Similarly in the case of gravitation. Newton described himself as unable to think that the attraction of one body for another at a distance, could be exerted in the absence of an intervening medium. But now let us ask how much the forwarder we are if an intervening medium be assumed. This ether whose undulations according to the received hypothesis constitute heat and light, and which is the vehicle of gravitation—how is it constituted? We must regard it, in the way that physicists do regard it, as composed of atoms which attract and repel each other—infinitesimal it may be in comparison with those of ordinary matter, but still atoms. And remembering that this ether is imponderable, we are obliged to conclude that the ratio between the interspaces of these atoms and the atoms themselves, is incommensurably greater than the like ratio in ponderable matter; else the densities could not be incommensurable. Instead then of a direct action by the Sun upon the Earth without anything intervening, we have to conceive the Sun's action propagated through a medium whose molecules are probably as small relatively to their interspaces as are the Sun and Earth compared with the space between them: we have to conceive these infinitesimal molecules acting on each other through absolutely vacant spaces which are immense in comparison with their own dimensions. How is this conception easier than the other? We still have mentally to represent a body as acting where it is not, and in the absence of anything by which its action may be transferred; and what matters it whether this takes place on a large or a small scale? We see therefore that the exercise of Force is altogether unintelligible. We cannot imagine it except through the instrumentality of something having extension; and yet when we have assumed this something, we find the perplexity is not got rid of but only postponed. We are obliged to conclude that matter, whether ponderable or imponderable, and whether aggregated or in its hypothetical units, acts upon matter through absolutely vacant space; and yet this conclusion is positively unthinkable.

Again, Light, Heat, Gravitation and all central forces, vary inversely as the squares of the distances; and physicists in their investigations assume that the units of matter act upon each other according to the same law—an assumption which indeed they are obliged to make; since this law is not simply an empirical one, but one deducible mathematically from the relations of space—one of which the negation is inconceivable. But now, in any mass of matter which is in internal equilibrium, what must follow? The attractions and repulsions of the constituent atoms are balanced. Being balanced, the atoms remain at their present distances; and the mass of matter neither expands nor contracts. But if the forces with which two adjacent atoms attract and repel each other both vary inversely as the squares of the distances, as they must; and if they are in equilibrium at their present distances, as they are; then, necessarily, they will be in equilibrium at all other distances. Let the atoms be twice as far apart, and their attractions and repulsions will both be reduced to one fourth of their present amounts. Let them be brought within half the distance, and their attractions and repulsions will both be quadrupled. Whence it follows that this matter will as readily as not assume any other density; and can offer no resistance to any external agents. Thus we are obliged to say that these antagonist molecular forces do not both vary inversely as the squares of the distances, which is unthinkable; or else that matter does not possess that attribute of resistance by which alone we distinguish it from empty space, which is absurd.

While then it is impossible to form any idea of Force in itself, it is equally impossible to comprehend either its mode of exercise or its law of variation.

§ 19. Turning now from the outer to the inner world, let us contemplate, not the agencies to which we ascribe our subjective modifications, but the subjective modifications themselves. These constitute a series. Difficult as we find it distinctly to separate and individualize them, it is nevertheless beyond question that our states of consciousness occur in succession.

Is this chain of states of consciousness infinite or finite? We cannot say infinite; not only because we have indirectly reached the conclusion that there was a period when it commenced, but also because all infinity is inconceivable—an infinite series included. We cannot say finite; for we have no knowledge of either of its ends. Go back in memory as far as we may, we are wholly unable to identify our first states of consciousness: the perspective of our thoughts vanishes in a dim obscurity where we can make out nothing. Similarly at the other extreme. We have no immediate knowledge of a termination to the series at a future time; and we cannot really lay hold of that temporary termination of the series reached at the present moment. For the state of consciousness recognized by us as our last, is not truly our last. That any mental affection may be contemplated as one of the series, it must be remembered—*represented* in thought, not *presented*. The truly last state of consciousness is that which is passing in the very act of contemplating a state just past—that in which we are thinking of the one before as the last. So that the proximate end of the chain eludes us, as well as the remote end.

"But," it may be said, "though we cannot directly *know* consciousness to be finite in duration, because neither of its limits can be actually reached; yet we can very well *conceive* it to be so." No: not even this is true. In the first place, we cannot *conceive* the terminations of that consciousness which alone we really know—our own—any more than we can *perceive* its terminations. For in truth the two acts are here one. In either case such terminations must be, as above said, not presented in thought, but represented; and they must be represented as in the act of occurring. Now to represent the termination of consciousness as occurring in ourselves, is to think of ourselves as contemplating the cessation of the last state of consciousness; and this implies a supposed continuance of consciousness after its last state, which is absurd. In the second place, if we regard the matter objectively—if we study the phenomena as occurring in others, or in the abstract, we are equally foiled. Consciousness implies perpetual change and the perpetual establishment of relations between its successive phases. To be known at all, any mental affection must be known as such or such—as like these foregoing ones or unlike those: if it is not thought of in connexion with others—not distinguished or identified by comparison with others, it is not recognized—is not a state of consciousness at all. A last state of consciousness, then, like any other, can exist only through a perception of its relations to previous states. But such perception of its relations must constitute a state later than the last, which is a contradiction. Or to put the difficulty in another form:—If ceaseless change of state is the condition on which alone consciousness exists, then when the supposed last state has been reached by the completion of the preceding change, change has ceased; therefore consciousness has ceased; therefore the supposed last state is not a state of consciousness at all; therefore there can be no last state of consciousness. In short, the perplexity is like that presented by the relations of Motion and Rest. As we found it was impossible really to conceive Rest becoming Motion or Motion becoming Rest; so here we find it is impossible really to conceive either the beginning or the ending of those changes which constitute consciousness.

Hence, while we are unable either to believe or to conceive that the duration of consciousness is infinite, we are equally unable either to know it as finite, or to conceive it as finite.

§ 20. Nor do we meet with any greater success when, instead of the extent of consciousness, we consider its substance. The question—What is this that thinks? admits of no better solution than the question to which we have just found none but inconceivable answers.

The existence of each individual as known to himself, has been always held by mankind at large, the most incontrovertible of truths. To say—"I am as sure of it as I am sure that I exist," is, in common speech, the most emphatic expression of certainty. And this fact of personal existence, testified to by the universal consciousness of men, has been made the basis of sundry philosophies; whence may be drawn the inference, that it is held by thinkers, as well as by the vulgar, to be beyond all facts unquestionable.

Belief in the reality of self, is, indeed, a belief which no hypothesis enables us to escape. What shall we say of these successive impressions and ideas which constitute consciousness? Shall we say that they are the affections of something called mind, which, as being the subject of them, is the real *ego*? If we say this, we manifestly imply that the *ego* is an entity. Shall we assert that these impressions and ideas are not the mere superficial changes wrought on some thinking substance, but are themselves the very body of this substance—are severally the modified forms which it from moment to moment assumes? This hypothesis, equally with the foregoing, implies that the individual exists as a permanent and distinct being; since modifications necessarily involve something modified. Shall we then betake ourselves to the sceptic's position, and argue that we know nothing more than our impressions and ideas themselves—that these are to us the only existences; and that the personality said to underlie them is a mere fiction? We do not even thus escape; since this proposition, verbally intelligible but really unthinkable, itself makes the assumption which it professes to repudiate. For how can consciousness be wholly resolved into impressions and ideas, when an impression of necessity implies something impressed? Or again, how can the sceptic who has decomposed his consciousness into impressions and ideas, explain the fact that he considers them as *his* impressions and ideas? Or once more, if, as he must, he admits that he has an impression of his personal existence, what warrant can he show for rejecting this impression as unreal while he accepts all his other impressions as real? Unless he can give satisfactory answers to these queries, which he cannot, he must abandon his conclusions; and must admit the reality of the individual mind.

But now, unavoidable as is this belief—established though it is, not only by the assent of mankind at large, endorsed by divers philosophers, but by the suicide of the sceptical argument—it is yet a belief admitting of no justification by reason: nay, indeed, it is a belief which reason, when pressed for a distinct answer, rejects. One of the most recent writers who has touched upon this question—Mr Mansel—does indeed contend that in the consciousness of self, we have a piece of real knowledge. The validity of immediate intuition he holds in this case unquestionable: remarking that "let system-makers say what they will, the unsophisticated sense of mankind refuses to acknowledge that mind is but a bundle of states of consciousness, as matter is (possibly) a bundle of sensible qualities." On which position the obvious comment is, that it does not seem altogether a consistent one for a Kantist, who pays but small respect to "the unsophisticated sense of mankind" when it testifies to the objectivity of space. Passing over this, however, it may readily be shown that a cognition of self, properly so called, is absolutely negatived by the laws of thought. The fundamental condition to all consciousness, emphatically insisted upon by Mr Mansel in common with Sir William Hamilton and others, is the antithesis of subject and object. And on this "primitive dualism of consciousness," "from which the explanations of philosophy must take their start," Mr Mansel founds his refutation of the German absolutists. But now, what is the corollary from this doctrine, as bearing on the consciousness of self? The mental act in which self is known, implies, like every other mental act, a perceiving subject and a perceived object. If, then, the object perceived is self, what is the subject that perceives? or if it is the true self which thinks, what other self can it be that is thought of? Clearly, a true cognition of self implies a state in which the knowing and the known are one—in which subject and object are identified; and this Mr Mansel rightly holds to be the annihilation of both.

So that the personality of which each is conscious, and of which the existence is to each a fact beyond all others the most certain, is yet a thing which cannot truly be known at all: knowledge of it is forbidden by the very nature of thought.

§ 21. Ultimate Scientific Ideas, then, are all representative of realities that cannot be comprehended. After no matter how great a progress in the colligation of facts and the establishment of generalizations ever wider and wider—after the merging of limited and derivative truths in truths that are larger and deeper has been carried no matter how far; the fundamental truth remains as much beyond reach as ever. The explanation of that which is explicable, does but bring out into greater clearness the inexplicableness of that which remains behind. Alike in the external and the internal worlds, the man of science sees himself in the midst of perpetual changes of which he can discover neither the beginning nor the end. If, tracing back the evolution of things, he allows himself to entertain the hypothesis that the Universe once existed in a diffused form, he finds it utterly impossible to conceive how this came to be so; and equally, if he speculates on the future, he can assign no limit to the grand succession of phenomena ever unfolding themselves before him. In like manner if he looks inward, he perceives that both ends of the thread of consciousness are beyond his grasp; nay, even beyond his power to think of as having existed or as existing in time to come. When, again, he turns from the succession of phenomena, external or internal, to their intrinsic nature, he is just as much at fault. Supposing him in every case able to resolve the appearances, properties, and movements of things, into manifestations of Force in Space and Time; he still finds that Force, Space, and Time pass all understanding. Similarly, though the analysis of mental actions may finally bring him down to sensations, as the original materials out of which all thought is woven, yet he is little forwarder; for he can give no account either of sensations themselves or of that something which is conscious of sensations. Objective and subjective things he thus ascertains to be alike inscrutable in their substance and genesis. In all directions his investigations eventually bring him face to face with an insoluble enigma; and he ever more clearly perceives it to be an insoluble enigma. He learns at once the greatness and the littleness of the human intellect—its power in dealing with all that comes within the range of experience; its impotence in dealing with all that transcends experience. He realizes with a special vividness the utter incomprehensibleness of the simplest fact, considered in itself. He, more than any other, truly *knows* that in its ultimate essence nothing can be known.

CHAPTER IV
THE RELATIVITY OF ALL KNOWLEDGE

§ 22. The same conclusion is thus arrived at, from whichever point we set out. If, respecting the origin and nature of things, we make some assumption, we find that through an inexorable logic it inevitably commits us to alternative impossibilities of thought; and this holds true of every assumption that can be imagined. If, contrariwise, we make no assumption, but set out from the sensible properties of surrounding objects, and, ascertaining their special laws of dependence, go on to merge these in laws more and more general, until we bring them all under some most general laws; we still find ourselves as far as ever from knowing what it is which manifests these properties to us: clearly as we seem to know it, our apparent knowledge proves on examination to be utterly irreconcilable with itself. Ultimate religious ideas and ultimate scientific ideas, alike turn out to be merely symbols of the actual, not cognitions of it.

The conviction, so reached, that human intelligence is incapable of absolute knowledge, is one that has been slowly gaining ground as civilization has advanced. Each new ontological theory, from time to time propounded in lieu of previous ones shown to be untenable, has been followed by a new criticism leading to a new scepticism. All possible conceptions have been one by one tried and found wanting; and so the entire field of speculation has been gradually exhausted without positive result: the only result arrived at being the negative one above stated—that the reality existing behind all appearances is, and must ever be, unknown. To this conclusion almost every thinker of note has subscribed. "With the exception," says Sir William Hamilton, "of a few late Absolutist theorisers in Germany, this is, perhaps, the truth of all others most harmoniously re-echoed by every philosopher of every school." And among these he names—Protagoras, Aristotle, St. Augustin, Boethius, Averroes, Albertus Magnus, Gerson, Leo Hebræus, Melancthon, Scaliger, Francis Piccolomini, Giordano Bruno, Campanella, Bacon, Spinoza, Newton, Kant.

It yet remains to point out how this belief may be established rationally, as well as empirically. Not only is it that, as in the earlier thinkers above named, a vague perception of the inscrutableness of things in themselves results from discovering the illusiveness of sense-impressions; and not only is it that, as shown in the foregoing chapters, definite experiments evolve alternative impossibilities of thought out of every ultimate conception we can frame; but it is that the relativity of our knowledge is demonstrable analytically. The induction drawn from general and special experiences, may be confirmed by a deduction from the nature of our intelligence. Two ways of reaching such a deduction exist. Proof that our cognitions are not, and never can be, absolute, is obtainable by analyzing either the *product* of thought, or the *process* of thought. Let us analyze each.

§ 23. If, when walking through the fields some day in September, you hear a rustle a few yards in advance, and on observing the ditch-side where it occurs, see the herbage agitated, you will probably turn towards the spot to learn by what this sound and motion are produced. As you approach there flutters into the ditch, a partridge; on seeing which your curiosity is satisfied—you have what you call an *explanation* of the appearances. The explanation, mark, amounts to this; that whereas throughout life you have had countless experiences of disturbance among small stationary bodies, accompanying the movement of other bodies among them, and have generalized the relation between such disturbances and such movements, you consider this particular disturbance explained, on finding it to present, an instance of the like relation. Suppose you catch the partridge; and, wishing to ascertain why it did not escape, examine it, and find at one spot, a slight trace of blood upon its feathers. You now *understand*, as you say, what has disabled the partridge. It has been wounded by a sportsman— adds another case to the many cases already seen by you, of birds being killed or injured by the shot discharged at them from fowling-pieces. And in assimilating this case to other such cases, consists your understanding of it. But now, on consideration, a difficulty suggests itself. Only a single shot has struck the partridge, and that not in a vital place: the wings are uninjured, as are also those muscles which move them; and the creature proves by its struggles that it has abundant strength. Why then, you inquire of yourself, does it not fly? Occasion favouring, you put the question to an anatomist, who furnishes you with *a solution.* He points out that this solitary shot has passed close to the place at which the nerve supplying the wing-muscles of one side, diverges from the spine; and that a slight injury to this nerve, extending even to the rupture of a few fibres, may, by preventing a perfect co-ordination in the actions of the two wings, destroy the power of flight. You are no longer puzzled. But what has happened?—what has changed your state from one of perplexity to one of *comprehension*? Simply the disclosure of a class of previously known cases, along with which you can include this case. The connexion between lesions of the nervous system and paralysis of limbs has been already many times brought under your notice; and you here find a relation of cause and effect that is essentially similar.

Let us suppose you are led on to make further inquiries concerning organic actions, which, conspicuous and remarkable as they are, you had not before cared to understand. How is respiration effected? you ask—why does air periodically rush into the lungs? The answer is that in the higher vertebrata, as in ourselves, influx of air is caused by an enlargement of the thoracic cavity, due, partly to depression of the diaphragm, partly to elevation of the ribs. But how does elevation of the ribs enlarge the cavity? In reply the anatomist shows you that the plane of each pair of ribs makes an acute angle with the spine; that this angle widens when the moveable ends of the ribs are raised; and he makes you realize the consequent dilatation of the cavity, by pointing out how the area of a parallelogram increases as its angles approach to right angles—you understand this special fact when you see it to be an instance of a general geometrical fact. There still arises, however, the question—why does the air rush into this enlarged cavity? To which comes the answer that, when the thoracic cavity is enlarged, the contained air, partially relieved from pressure, expands, and so loses some of its resisting power; that hence it opposes to the pressure of the external air a less pressure; and that as air, like every other fluid, presses equally in all directions, motion must result along any line in which the resistance is less than elsewhere; whence follows an inward current. And this *interpretation* you recognize as one, when a few facts of like kind, exhibited more plainly in a visible fluid such as water, are cited in illustration. Again, when it was pointed out that the limbs are compound levers acting in essentially the same way as levers of iron or wood, you might consider yourself as having obtained a partial *rationale* of animal movements. The contraction of a muscle, seeming before utterly unaccountable, would seem less unaccountable were you shown how, by a galvanic current, a series of soft iron magnets could be made to shorten itself, through the attraction of each magnet for its neighbours:—an alleged analogy which especially answers the purpose of our argument; since, whether real or fancied, it equally illustrates the mental illumination that results on finding a class of cases within which a particular case may possibly be included. And it may be further noted how, in the instance here named, an additional feeling of comprehension arises on remembering that the influence conveyed through the nerves to the muscles, is, though not positively electric, yet a form of force nearly allied to the electric. Similarly when you learn that animal heat arises from chemical combination, and so is evolved as heat is evolved in other chemical combinations—when you learn that the absorption of nutrient fluids through the coats of the intestines, is an instance of osmotic action—when you learn that the changes undergone by food during digestion, are like changes artificially producible in the laboratory; you regard yourself as *knowing* something about the natures of these phenomena.

Observe now what we have been doing. Turning to the general question, let us note where these successive interpretations have carried us. We began with quite special and concrete facts. In explaining each, and afterwards explaining the more general facts of which they are instances, we have got down to certain highly general facts: —to a geometrical principle or property of space, to a simple law of mechanical action, to a law of fluid equilibrium—to truths in physics, in chemistry, in thermology, in electricity. The particular phenomena with which we set out, have been merged in larger and larger groups of phenomena; and as they have been so merged, we have arrived at solutions that we consider profound in proportion as this process has been carried far. Still deeper explanations are simply further steps in the same direction. When, for instance, it is asked why the law of action of the lever is what it is, or why fluid equilibrium and fluid motion exhibit the relations which they do, the answer furnished by mathematicians consists in the disclosure of the principle of virtual velocities—a principle holding true alike in fluids and solids—a principle under which the others are comprehended. And similarly, the insight obtained into the phenomena of chemical combination, heat, electricity, &c., implies that a rationale of them, when found, will be the exposition of some highly general fact respecting the constitution of matter, of which chemical, electrical, and thermal facts, are merely different manifestations.

Is this process limited or unlimited? Can we go on for ever explaining classes of facts by including them in larger classes; or must we eventually come to a largest class? The supposition that the process is unlimited, were any one absurd enough to espouse it, would still imply that an ultimate explanation could not be reached; since infinite time would be required to reach it. While the unavoidable conclusion that it is limited (proved not only by the finite sphere of observation open to us, but also by the diminution in the number of generalizations that necessarily accompanies increase of their breadth) equally implies that the ultimate fact cannot be understood. For if the successively deeper interpretations of nature which constitute advancing knowledge, are merely successive inclusions of special truths in general truths, and of general truths in truths still more general; it obviously follows that the most general truth, not admitting of inclusion in any other, does not admit of interpretation. Manifestly, as the *most* general cognition at which we arrive cannot be reduced to a *more* general one, it cannot be understood. Of necessity, therefore, explanation must eventually bring us down to the inexplicable. The deepest truth which we can get at, must be unaccountable. Comprehension must become something other than comprehension, before the ultimate fact can be comprehended.

§ 24. The inference which we thus find forced upon us when we analyze the product of thought, as exhibited objectively in scientific generalizations, is equally forced upon us by an analysis of the process of thought, as exhibited subjectively in consciousness. The demonstration of the necessarily relative character of our knowledge, as deduced from the nature of intelligence, has been brought to its most definite shape by Sir William Hamilton. I cannot here do better than extract from his essay on the "Philosophy of the Unconditioned," the passage containing the substance of his doctrine.

"The mind can conceive," he argues, "and consequently can know," only the *limited, and the conditionally limited.* The unconditionally unlimited, or the *Infinite,* the unconditionally limited, or the *Absolute,* cannot positively be construed to the mind; they can be conceived, only by a thinking away from, or abstraction of, those very conditions under which thought itself is realized; consequently, the notion of the Unconditioned is only negative,— negative of the conceivable itself. For example, on the one hand we can positively conceive, neither an absolute whole, that is, a whole so great, that we cannot also conceive it as a relative part of a still greater whole; nor an absolute part, that is, a part so small, that we cannot also conceive it as a relative whole, divisible into smaller parts. On the other hand, we cannot positively represent, or realize, or construe to the mind (as here understanding and imagination coincide), an infinite whole, for this could only be done by the infinite synthesis in thought of finite wholes, which would itself require an infinite time for its accomplishment; nor, for the same reason, can we follow out in thought an infinite divisibility of parts. The result is the same, whether we apply the process to limitation in *space,* in *time,* or in *degree.* The unconditional negation, and the unconditional affirmation of limitation; in other words, the *infinite* and *absolute, properly so called,* are thus equally inconceivable to us.

As the conditionally limited (which we may briefly call the *conditioned*) is thus the only possible object of knowledge and of positive thought—thought necessarily supposes conditions. To *think* is to *condition*; and conditional limitation is the fundamental law of the possibility of thought. For, as the greyhound cannot outstrip his shadow, nor (by a more appropriate simile) the eagle outsoar the atmosphere in which he floats, and by which alone he may be supported; so the mind cannot transcend that sphere of limitation, within and through which exclusively the possibility of thought is realized. Thought is only of the conditioned; because, as we have said, to think is simply to condition. The *absolute* is conceived merely by a negation of conceivability; and all that we know, is only known as

——'won from the void and formless *infinite.*'

How, indeed, it could ever be doubted that thought is only of the conditioned, may well be deemed a matter of the profoundest admiration. Thought cannot transcend consciousness; consciousness is only possible under the antithesis of a subject and object of thought, known only in correlation, and mutually limiting each other; while, independently of this, all that we know either of subject or object, either of mind or matter, is only a knowledge in each of the particular, of the plural, of the different, of the modified, of the phenomenal. We admit that the consequence of this doctrine is,—that philosophy, if viewed as more than a science of the conditioned, is impossible. Departing from the particular, we admit, that we can never, in our highest generalizations, rise above the finite; that our knowledge, whether of mind or matter, can be nothing more than a knowledge of the relative manifestations of an existence, which in itself it is our highest wisdom to recognize as beyond the reach of philosophy,—in the language of St Austin,—'*cognoscendo ignorari, et ignorando cognosci.*'

"The conditioned is the mean between two extremes,—two inconditionates, exclusive of each other, neither of which *can be conceived as possible*, but of which, on the principles of contradiction and excluded middle, one *must be admitted as necessary.* On this opinion, therefore, reason is shown to be weak, but not deceitful. The mind is not represented as conceiving two propositions subversive of each other, as equally possible; but only, as unable to understand as possible, either of two extremes; one of which, however, on the ground of their mutual repugnance, it is compelled to recognize as true. We are thus taught the salutary lesson, that the capacity of thought is not to be constituted into the measure of existence; and are warned from recognizing the domain of our knowledge as necessarily co-extensive with the horizon of our faith. And by a wonderful revelation, we are thus, in the very consciousness of our inability to conceive aught above the relative and finite, inspired with a belief in the existence of something unconditioned beyond the sphere of all comprehensible reality."

Clear and conclusive as this statement of the case appears when carefully studied, it is expressed in so abstract a manner as to be not very intelligible to the general reader. A more popular presentation of it, with illustrative applications, as given by Mr Mansel in his "Limits of Religious Thought," will make it more fully understood. The following extracts, which I take the liberty of making from his pages, will suffice.

"The very conception of consciousness, in whatever mode it may be manifested, necessarily implies *distinction between one object and another*. To be conscious, we must be conscious of something; and that something can only be known, as that which it is, by being distinguished from that which it is not. But distinction is necessarily limitation; for, if one object is to be distinguished from another, it must possess some form of existence which the other has not, or it must not possess some form which the other has. But it is obvious the Infinite cannot be distinguished, as such, from the Finite, by the absence of any quality which the Finite possesses; for such absence would be a limitation. Nor yet can it be distinguished by the presence of an attribute which the Finite has not; for, as no finite part can be a constituent of an infinite whole, this differential characteristic must itself be infinite; and must at the same time have nothing in common with the finite. We are thus thrown back upon our former impossibility; for this second infinite will be distinguished from the finite by the absence of qualities which the latter possesses. A consciousness of the Infinite as such thus necessarily involves a self-contradiction; for it implies the recognition, by limitation and difference, of that which can only be given as unlimited and indifferent.

"This contradiction, which is utterly inexplicable on the supposition that the infinite is a positive object of human thought, is at once accounted for, when it is regarded as the mere negation of thought. If all thought is limitation;—if whatever we conceive is, by the very act of conception, regarded as finite,—*the infinite*, from a human point of view, is merely a name for the absence of those conditions under which thought is possible. To speak of a *Conception of the Infinite* is, therefore, at once to affirm those conditions and to deny them. The contradiction, which we discover in such a conception, is only that which we have ourselves placed there, by tacitly assuming the conceivability of the inconceivable. The condition of consciousness is distinction; and condition of distinction is limitation. We can have no consciousness of Being in general which is not some Being in particular: a *thing*, in consciousness, is one thing out of many. In assuming the possibility of an infinite object of consciousness, I assume, therefore, that it is at the same time limited and unlimited;—actually something, without which it could not be an object of consciousness, and actually nothing, without which it could not be infinite.

"A second characteristic of Consciousness is, that it is only possible in the form of a *relation*. There must be a Subject, or person conscious, and an Object, or thing of which he is conscious. There can be no consciousness without the union of these two factors; and, in that union, each exists only as it is related to the other. The subject is a subject, only in so far as it is conscious of an object: the object is an object, only in so far as it is apprehended by a subject: and the destruction of either is the destruction of consciousness itself. It is thus manifest that a consciousness of the Absolute is equally self-contradictory with that of the Infinite. To be conscious of the Absolute as such, we must know that an object, which is given in relation to our consciousness, is identical with one which exists in its own nature, out of all relation to consciousness. But to know this identity, we must be able to compare the two together; and such a comparison is itself a contradiction. We are in fact required to compare that of which we are conscious with that of which we are not conscious; the comparison itself being an act of consciousness, and only possible through the consciousness of both its objects. It is thus manifest that, even if we could be conscious of the absolute, we could not possibly know that it is the absolute: and, as we can be conscious of an object as such, only by knowing it to be what it is, this is equivalent to an admission that we cannot be conscious of the absolute at all. As an object of consciousness, every thing is necessarily relative; and what a thing may be out of consciousness, no mode of consciousness can tell us.

"This contradiction, again, admits of the same explanation as the former. Our whole notion of existence is necessarily relative; for it is existence as conceived by us. But *Existence*, as we conceive it, is but a name for the several ways in which objects are presented to our consciousness,—a general term, embracing a variety of relations. *The Absolute*, on the other hand, is a term expressing no object of thought, but only a denial of the relation by which thought is constituted. To assume absolute existence as an object of thought, is thus to suppose a relation existing when the related terms exist no longer. An object of thought exists, as such, in and through its relation to a thinker; while the Absolute, as such, is independent of all relation. The *Conception of the Absolute* thus implies at the same time the presence and absence of the relation by which thought is constituted; and our various endeavours to represent it are only so many modified forms of the contradiction involved in our original assumption. Here, too, the contradiction is one which we ourselves have made. It does not imply that the Absolute cannot exist; but it implies, most certainly, that we cannot conceive it as existing."

Here let me point out how the same general inference may be evolved from another fundamental condition of thought, omitted by Sir W. Hamilton, and not supplied by Mr Mansel;—a condition which, under its obverse aspect, we have already contemplated in the last section. Every complete act of consciousness, besides distinction and relation, also implies likeness. Before it can become an idea, or constitute a piece of knowledge, a mental state must not only be known as separate in kind from certain foregoing states to which it is known as related by succession; but it must further be known as of the same kind with certain other foregoing states. That organization of changes which constitutes thinking, involves continuous integration as well as continuous differentiation. Were each new affection of the mind perceived simply as an affection in some way contrasted with the preceding ones—were there but a chain of impressions, each of which as it arose was merely distinguished from its predecessors; consciousness would be an utter chaos. To produce that orderly consciousness which we call intelligence, there requires the assimilation of each impression to others, that occurred earlier in the series. Both the successive mental states, and the successive relations which they bear to each other, must be classified; and classification involves not only a parting of the unlike, but also a binding together of the like. In brief, a true cognition is possible only through an accompanying recognition. Should it be objected that if so, there cannot be a first cognition, and hence there can be no cognition; the reply is, that cognition proper arises gradually—that during the first stage of incipient intelligence, before the feelings produced by intercourse with the outer world have been put into order, there *are* no cognitions, strictly so called; and that, as every infant shows us, these slowly emerge out of the confusion of unfolding consciousness as fast as the experiences are arranged into groups—as fast as the most frequently repeated sensations, and their relations to each other, become familiar enough to admit of their recognition as such or such, whenever they recur. Should it be further objected that if cognition pre-supposes recognition, there can be, no cognition, even by an adult, of an object never before seen; there is still the sufficient answer that in so far as it is not assimilated to previously-seen objects, it is *not* known, and that it *is* known in so far as it is assimilated to them. Of this paradox the interpretation is, that an object is classifiable in various ways, with various degrees of completeness. An animal hitherto *unknown* (mark the word), though not referable to any established species or genus, is yet *recognized* as belonging to one of the larger divisions—mammals, birds, reptiles, or fishes; or should it be so anomalous that its alliance with any of these is not determinable, it may yet be classed as vertebrate or invertebrate; or if it be one of those organisms of which it is doubtful whether the animal or vegetal characteristics predominate, it is still known as a living body; even should it be questioned whether it is organic, it remains beyond question that it is a material object, and it is cognized by being recognized as such. Whence it is manifest that a thing is perfectly known only when it is in all respects like certain things previously observed; that in proportion to the number of respects in which it is unlike them, is the extent to which it is unknown; and that hence when it has absolutely no attribute in common with anything else, it must be absolutely beyond the bounds of knowledge.

Observe the corollary which here concerns us. A cognition of the Real, as distinguished from the Phenomenal, must, if it exists, conform to this law of cognition in general. The First Cause, the Infinite, the Absolute, to be known at all, must be classed. To be positively thought of, it must be thought of as such or such—as of this or that kind. Can it be like in kind to anything of which we have sensible experience? Obviously not. Between the creating and the created, there must be a distinction transcending any of the distinctions existing between different divisions of the created. That which is uncaused cannot be assimilated to that which is caused: the two being, in the very naming, antithetically opposed. The Infinite cannot be grouped along with something that is finite; since, in being so grouped, it must be regarded as not-infinite. It is impossible to put the Absolute in the same category with anything relative, so long as the Absolute is defined as that of which no necessary relation can be predicated. Is it then that the Actual, though unthinkable by classification with the Apparent, is thinkable by classification with itself? This supposition is equally absurd with the other. It implies the plurality of the First Cause, the Infinite, the Absolute; and this implication is self-contradictory. There cannot be more than one First Cause; seeing that the existence of more than one would involve the existence of something necessitating more than one, which something would be the true First Cause. How self-destructive is the assumption of two or more Infinites, is manifest on remembering that such Infinites, by limiting each other, would become finite. And similarly, an Absolute which existed not alone but along with other Absolutes, would no longer be an absolute but a relative. The Unconditioned therefore, as class-able neither with any form of the conditioned nor with any other Unconditioned, cannot be classed at all. And to admit that it cannot be known as of such or such kind, is to admit that it is unknowable.

Thus, from the very nature of thought, the relativity of our knowledge is inferable in three several ways. As we find by analyzing it, and as we see it objectively displayed in every proposition, a thought involves *relation, difference, likeness.* Whatever does not present each of these does not admit of cognition. And hence we may say that the Unconditioned, as presenting none of them, is trebly unthinkable.

§ 25. From yet another point of view we may discern the same great truth. If, instead of examining our intellectual powers directly as exhibited in the act of thought, or indirectly as exhibited in thought when expressed by words, we look at the connexion between the mind and the world, a like conclusion is forced upon us. In the very definition of Life, when reduced to its most abstract shape, this ultimate implication becomes visible.

All vital actions, considered not separately but in their ensemble, have for their final purpose the balancing of certain outer processes by certain inner processes. There are unceasing external forces tending to bring the matter of which organic bodies consist, into that state of stable equilibrium displayed by inorganic bodies; there are internal forces by which this tendency is constantly antagonized; and the perpetual changes which constitute Life, may be regarded as incidental to the maintenance of the antagonism. To preserve the erect posture, for instance, we see that certain weights have to be neutralized by certain strains: each limb or other organ, gravitating to the Earth and pulling down the parts to which it is attached, has to be preserved in position by the tension of sundry muscles; or in other words, the group of forces which would if allowed bring the body to the ground, has to be counterbalanced by another group of forces. Again, to keep up the temperature at a particular point, the external process of radiation and absorption of heat by the surrounding medium, must be met by a corresponding internal process of chemical combination, whereby more heat may be evolved; to which add, that if from atmospheric changes the loss becomes greater or less, the production must become greater or less. And similarly throughout the organic actions in general.

When we contemplate the lower kinds of life, we see that the correspondences thus maintained are direct and simple; as in a plant, the vitality of which mainly consists in osmotic and chemical actions responding to the co-existence of light, heat, water, and carbonic acid around it. But in animals, and especially in the higher orders of them, the correspondences become extremely complex. Materials for growth and repair not being, like those which plants require, everywhere present, but being widely dispersed and under special forms, have to be found, to be secured, and to be reduced to a fit state for assimilation. Hence the need for locomotion; hence the need for the senses; hence the need for prehensile and destructive appliances; hence the need for an elaborate digestive apparatus. Observe, however, that these successive complications are essentially nothing but aids to the maintenance of the organic balance in its integrity, in opposition to those physical, chemical, and other agencies which tend to overturn it. And observe, moreover, that while these successive complications subserve this fundamental adaptation of inner to outer actions, they are themselves nothing else but further adaptations of inner to outer actions. For what are those movements by which a predatory creature pursues its prey, or by which its prey seeks to escape, but certain changes in the organism fitted to meet certain changes in its environment? What is that compound operation which constitutes the perception of a piece of food, but a particular correlation of nervous modifications, answering to a particular correlation of physical properties? What is that process by which food when swallowed is reduced to a fit form for assimilation, but a set of mechanical and chemical actions responding to the mechanical and chemical actions which distinguish the food? Whence it becomes manifest, that while Life in its simplest form is the correspondence of certain inner physico-chemical actions with certain outer physico-chemical actions, each advance to a higher form of Life consists in a better preservation of this primary correspondence by the establishment of other correspondences.

Divesting this conception of all superfluities and reducing it to its most abstract shape, we see that Life is definable as the continuous adjustment of internal relations to external relations. And when we so define it, we discover that the physical and the psychial life are equally comprehended by the definition. We perceive that this which we call Intelligence, shows itself when the external relations to which the internal ones are adjusted, begin to be numerous, complex, and remote in time or space; that every advance in Intelligence essentially consists in the establishment of more varied, more complete, and more involved adjustments; and that even the highest achievements of science are resolvable into mental relations of co-existence and sequence, so co-ordinated as exactly to tally with certain relations of co-existence and sequence that occur externally. A caterpillar, wandering at random and at length finding its way on to a plant having a certain odour, begins to eat—has inside of it an organic relation between a particular impression and a particular set of actions, answering to the relation outside of it, between scent and nutriment. The sparrow, guided by the more complex correlation of impressions which the colour, form, and movements of the caterpillar gave it; and guided also by other correlations which measure the position and distance of the caterpillar; adjusts certain correlated muscular movements in such way as to seize the caterpillar. Through a much greater distance in space is the hawk, hovering above, affected by the relations of shape and motion which the sparrow presents; and the much more complicated and prolonged series of related nervous and muscular changes, gone through in correspondence with the sparrow's changing relations of position, finally succeed when they are precisely adjusted to these changing relations. In the fowler, experience has established a relation between the appearance and flight of a hawk and the destruction of other birds, including game; there is also in him an established relation between those visual impressions answering to a certain distance in space, and the range of his gun; and he has learned, too, by frequent observation, what relations of position the sights must bear to a point somewhat in advance of the flying bird, before he can fire with success. Similarly if we go back to the manufacture of the gun. By relations of co-existence between colour, density, and place in the earth, a particular mineral is known as one which yields iron; and the obtainment of iron from it, results when certain correlated acts of ours, are adjusted to certain correlated affinities displayed by ironstone, coal, and lime, at a high temperature. If we descend yet a step further, and ask a chemist to explain the explosion of gunpowder, or apply to a mathematician for a theory of projectiles, we still find that special or general relations of co-existence and sequence between properties, motions, spaces &c., are all they can teach us. And lastly, let it be noted that what we call *truth*, guiding us to successful action and the consequent maintenance of life, is simply the accurate correspondence of subjective to objective relations; while *error*, leading to failure and therefore towards death, is the absence of such accurate correspondence.

If, then, Life in all its manifestations, inclusive of Intelligence in its highest forms, consists in the continuous adjustment of internal relations to external relations, the necessarily relative character of our knowledge becomes obvious. The simplest cognition being the establishment of some connexion between subjective states, answering to some connexion between objective agencies; and each successively more complex cognition being the establishment of some more involved connexion of such states, answering to some more involved connexion of such agencies; it is clear that the process, no matter how far it be carried, can never bring within the reach of Intelligence, either the states themselves or the agencies themselves. Ascertaining which things occur along with which, and what things follow what, supposing it to be pursued exhaustively, must still leave us with co-existences and sequences only. If every act of knowing is the formation of a relation in consciousness parallel to a relation in the environment, then the relativity of knowledge is self-evident—becomes indeed a truism. Thinking being relationing, no thought can ever express more than relations.

And here let us not omit to mark how that to which our intelligence is confined, is that with which alone our intelligence is concerned. The knowledge within our reach, is the only knowledge that can be of service to us. This maintenance of a correspondence between internal actions and external actions, which both constitutes our life at each moment and is the means whereby life is continued through subsequent moments, merely requires that the agencies acting upon us shall be known in their co-existences and sequences, and not that they shall be known in themselves. If x and y are two uniformly connected properties in some outer object, while a and b are the effects they produce in our consciousness; and if while the property x produces in us the indifferent mental state a, the property y produces in us the painful mental state b (answering to a physical injury); then, all that is requisite for our guidance, is, that x being the uniform accompaniment of y externally, a shall be the uniform accompaniment of b internally; so that when, by the presence of x, a is produced in consciousness, b, or rather the idea of b, shall follow it, and excite the motions by which the effect of y may be escaped. The sole need is that a and b and the relation between them, shall always answer to x and y and the relation between them. It matters nothing to us if a and b are like x and y or not. Could they be exactly identical with them, we should not be one whit the better off; and their total dissimilarity is no disadvantage to us.

Deep down then in the very nature of Life, the relativity of our knowledge is discernible. The analysis of vital actions in general, leads not only to the conclusion that things in themselves cannot be known to us; but also to the conclusion that knowledge of them, were it possible, would be useless.

§ 26. There still remains the final question—What must we say concerning that which transcends knowledge? Are we to rest wholly in the consciousness of phenomena?—is the result of inquiry to exclude utterly from our minds everything but the relative? or must we also believe in something beyond the relative?

The answer of pure logic is held to be, that by the limits of our intelligence we are rigorously confined within the relative; and that anything transcending the relative can be thought of only as a pure negation, or as a non-existence. "The *absolute* is conceived merely by a negation of conceivability," writes Sir William Hamilton. "The *Absolute* and the *Infinite*" says Mr Mansel, "are thus, like the *Inconceivable* and the *Imperceptible*, names indicating, not an object of thought or of consciousness at all, but the mere absence of the conditions under which consciousness is possible." From each of which extracts may be deduced the conclusion, that since reason cannot warrant us in affirming the positive existence of what is cognizable only as a negation, we cannot rationally affirm the positive existence of anything beyond phenomena.

Unavoidable as this conclusion seems, it involves, I think, a grave error. If the premiss be granted, the inference must doubtless be admitted; but the premiss, in the form presented by Sir William Hamilton and Mr Mansel, is not strictly true. Though, in the foregoing pages, the arguments used by these writers to show that the Absolute is unknowable, have been approvingly quoted; and though these arguments have been enforced by others equally thoroughgoing; yet there remains to be stated a qualification, which saves us from that scepticism otherwise necessitated. It is not to be denied that so long as we confine ourselves to the purely logical aspect of the question, the propositions quoted above must be accepted in their entirety; but when we contemplate its more general, or psychological, aspect, we find that these propositions are imperfect statements of the truth: omitting, or rather excluding, as they do, an all-important fact. To speak specifically:—Besides that *definite* consciousness of which Logic formulates the laws, there is also an *indefinite* consciousness which cannot be formulated. Besides complete thoughts, and besides the thoughts which though incomplete admit of completion, there are thoughts which it is impossible to complete; and yet which are still real, in the sense that they are normal affections of the intellect.

Observe in the first place, that every one of the arguments by which the relativity of our knowledge is demonstrated, distinctly postulates the positive existence of something beyond the relative. To say that we cannot know the Absolute, is, by implication, to affirm that there *is* an Absolute. In the very denial of our power to learn *what* the Absolute is, there lies hidden the assumption *that* it is; and the making of this assumption proves that the Absolute has been present to the mind, not as a nothing, but as a something. Similarly with every step in the reasoning by which this doctrine is upheld. The Noumenon, everywhere named as the antithesis of the Phenomenon, is throughout necessarily thought of as an actuality. It is rigorously impossible to conceive that our knowledge is a knowledge of Appearances only, without at the same time conceiving a Reality of which they are appearances; for appearance without reality is unthinkable. Strike out from the argument the terms Unconditioned, Infinite, Absolute, with their equivalents, and in place of them write, "negation of conceivability," or "absence of the conditions under which consciousness is possible," and you find that the argument becomes nonsense. Truly to realize in thought any one of the propositions of which the argument consists, the Unconditioned must be represented as positive and not negative. How then can it be a legitimate conclusion from the argument, that our consciousness of it is negative? An argument, the very construction of which assigns to a certain term a certain meaning, but which ends in showing that this term has no such meaning, is simply an elaborate suicide. Clearly, then, the very demonstration that a *definite* consciousness of the Absolute is impossible to us, unavoidably presupposes an *indefinite* consciousness of it.

Perhaps the best way of showing that by the necessary conditions of thought, we are obliged to form a positive though vague consciousness of this which transcends distinct consciousness, is to analyze our conception of the antithesis between Relative and Absolute. It is a doctrine called in question by none, that such antinomies of thought as Whole and Part, Equal and Unequal, Singular and Plural, are necessarily conceived as correlatives: the conception of a part is impossible without the conception of a whole; there can be no idea of equality without one of inequality. And it is admitted that in the same manner, the Relative is itself conceivable as such, only by opposition to the Irrelative or Absolute. Sir William Hamilton however, in his trenchant (and in most parts unanswerable) criticism on Cousin, contends, in conformity with his position above stated, that one of these correlatives is nothing whatever beyond the negation of the other. "Correlatives" he says "certainly suggest each other, but correlatives may, or may not, be equally real and positive. In thought contradictories necessarily imply each other, for the knowledge of contradictories is one. But the reality of one contradictory, so far from guaranteeing the reality of the other, is nothing else than its negation. Thus every positive notion (the concept of a thing by what it is) suggests a negative notion (the concept of a thing by what it is not); and the highest positive notion, the notion of the conceivable, is not without its corresponding negative in the notion of the inconceivable. But though these mutually suggest each other, the positive alone is real; the negative is only an abstraction of the other, and in the highest generality, even an abstraction of thought itself." Now the assertion that of such contradictories "the negative is *only* an abstraction of the other" — "is *nothing* else than its negation," — is not true. In such correlatives as Equal and Unequal, it is obvious enough that the negative concept contains something besides the negation of the positive one; for the things of which equality is denied are not abolished from consciousness by the denial. And the fact overlooked by Sir William Hamilton, is, that the like holds even with those correlatives of which the negative is inconceivable, in the strict sense of the word. Take for example the Limited and the Unlimited. Our notion of the Limited is composed, firstly of a consciousness of some kind of being, and secondly of a consciousness of the limits under which it is known. In the antithetical notion of the Unlimited, the consciousness of limits is abolished; but not the consciousness of some kind of being. It is quite true that in the absence of conceived limits, this consciousness ceases to be a concept properly so called; but it is none the less true that it remains as a mode of consciousness. If, in such cases, the negative contradictory were, as alleged, *"nothing else"* than the negation of the other, and therefore a mere nonentity, then it would clearly follow that negative contradictories could be used interchangeably: the Unlimited might be thought of as antithetical to the Divisible; and the Indivisible as antithetical to the Limited. While the fact that they cannot be so used, proves that in consciousness the Unlimited and the Indivisible are qualitatively distinct, and therefore positive or real; since distinction cannot exist between nothings. The error, (very naturally fallen into by philosophers intent on demonstrating the limits and conditions of consciousness,) consists in assuming that consciousness contains *nothing but* limits and conditions; to the entire neglect of that which is limited and conditioned. It is forgotten that there is something which alike forms the raw material of definite thought and remains after the definiteness which thinking gave to it has been destroyed. Now all this applies by change of terms to the last and highest of these antinomies—that between the Relative and the Non-relative. We are conscious of the Relative as existence under conditions and limits; it is impossible that these conditions and limits can be thought of apart from something to which they give the form; the abstraction of these conditions and limits, is, by the hypothesis, the abstraction of them *only*; consequently there must be a residuary consciousness of something which filled up their outlines; and this indefinite something constitutes our consciousness of the Non-relative or Absolute. Impossible though it is to give to this consciousness any qualitative or quantitative expression whatever, it is not the less certain that it remains with us as a positive and indestructible element of thought.

Still more manifest will this truth become when it is observed that our conception of the Relative itself disappears, if our conception of the Absolute is a pure negation. It is admitted, or rather it is contended, by the writers I have quoted above, that contradictories can be known only in relation to each other—that Equality, for instance, is unthinkable apart from its correlative Inequality; and that thus the Relative can itself be conceived only by opposition to the Non-relative. It is also admitted, or rather contended, that the consciousness of a relation implies a consciousness of both the related members. If we are required to conceive the relation between the Relative and Non-relative without being conscious of both, "we are in fact" (to quote the words of Mr Mansel differently applied) "required to compare that of which we are conscious with that of which we are not conscious; the comparison itself being an act of consciousness, and only possible through the consciousness of both its objects." What then becomes of the assertion that "the Absolute is conceived merely by a negation of conceivability," or as "the mere absence of the conditions under which consciousness is possible?" If the Non-relative or Absolute, is present in thought only as a mere negation, then the relation between it and the Relative becomes unthinkable, because one of the terms of the relation is absent from consciousness. And if this relation is unthinkable, then is the Relative itself unthinkable, for want of its antithesis: whence results the disappearance of all thought whatever.

Let me here point out that both Sir Wm Hamilton and Mr Mansel, do, in other places, distinctly imply that our consciousness of the Absolute, indefinite though it is, is positive and not negative. The very passage already quoted from Sir Wm Hamilton, in which he asserts that "the *absolute* is conceived merely by a negation of conceivability," itself ends with the remark that, "by a wonderful revelation, we are thus, in the very consciousness of our inability to conceive aught above the relative and finite, inspired with a belief in the existence of something unconditioned beyond the sphere of all comprehensible reality." The last of these assertions practically admits that which the other denies. By the laws of thought as Sir Wm Hamilton has interpreted them, he finds himself forced to the conclusion that our consciousness of the Absolute is a pure negation. He nevertheless finds that there does exist in consciousness an irresistible conviction of the real "existence of something unconditioned." And he gets over the inconsistency by speaking of this conviction as "a wonderful revelation"—"a belief" with which we are "inspired:" thus apparently hinting that it is supernaturally at variance with the laws of thought. Mr Mansel is betrayed into a like inconsistency. When he says that "we are compelled, by the constitution of our minds, to believe in the existence of an Absolute and Infinite Being,—a belief which appears forced upon us, as the complement of our consciousness of the relative and the finite;" he clearly says by implication that this consciousness is positive, and not negative. He tacitly admits that we are obliged to regard the Absolute as something more than a negation—that our consciousness of it is *not* "the mere absence of the conditions under which consciousness is possible."

The supreme importance of this question must be my apology for taxing the reader's attention a little further, in the hope of clearing up the remaining difficulties. The necessarily positive character of our consciousness of the Unconditioned, which, as we have seen, follows from an ultimate law of thought, will be better understood on contemplating the process of thought.

One of the arguments used to prove the relativity of our knowledge, is, that we cannot conceive Space or Time as either limited or unlimited. It is pointed out that when we imagine a limit, there simultaneously arises the consciousness of a space or time existing beyond the limit. This remoter space or time, though not contemplated as definite, is yet contemplated as real. Though we do not form of it a conception proper, since we do not bring it within bounds, there is yet in our minds the unshaped material of a conception. Similarly with our consciousness of Cause. We are no more able to form a circumscribed idea of Cause, than of Space or Time; and we are consequently obliged to think of the Cause which transcends the limits of our thought as positive though indefinite. Just in the same manner that on conceiving any bounded space, there arises a nascent consciousness of space outside the bounds; so, when we think of any definite cause, there arises a nascent consciousness of a cause behind it: and in the one case as in the other, this nascent consciousness is in substance like that which suggests it, though without form. The momentum of thought inevitably carries us beyond conditioned existence to unconditioned existence; and this ever persists in us as the body of a thought to which we can give no shape.

Hence our firm belief in objective reality—a belief which metaphysical criticisms cannot for a moment shake. When we are taught that a piece of matter, regarded by us as existing externally, cannot be really known, but that we can know only certain impressions produced on us, we are yet, by the relativity of our thought, compelled to think of these in relation to a positive cause—the notion of a real existence which generated these impressions becomes nascent. If it be proved to us that every notion of a real existence which we can frame, is utterly inconsistent with itself—that matter, however conceived by us, cannot be matter as it actually is, our conception, though transfigured, is not destroyed: there remains the sense of reality, dissociated as far as possible from those special forms under which it was before represented in thought. Though Philosophy condemns successively each attempted conception of the Absolute—though it proves to us that the Absolute is not this, nor that, nor that—though in obedience to it we negative, one after another, each idea as it arises; yet, as we cannot expel the entire contents of consciousness, there ever remains behind an element which passes into new shapes. The continual negation of each particular form and limit, simply results in the more or less complete abstraction of all forms and limits; and so ends in an indefinite consciousness of the unformed and unlimited.

And here we come face to face with the ultimate difficulty—How can there possibly be constituted a consciousness of the unformed and unlimited, when, by its very nature, consciousness is possible only under forms and limits? If every consciousness of existence is a consciousness of existence as conditioned, then how, after the negation of conditions, can there be any residuum?. Though not directly withdrawn by the withdrawal of its conditions, must not the raw material of consciousness be withdrawn by implication? Must it not vanish when the conditions of its existence vanish? That there must be a solution of this difficulty is manifest; since even those who would put it, do, as already shown, admit that we have some such consciousness; and the solution appears to be that above shadowed forth. Such consciousness is not, and cannot be, constituted by any single mental act; but is the product of many mental acts. In each concept there is an element which persists. It is alike impossible for this element to be absent from consciousness, and for it to be present in consciousness alone: either alternative involves unconsciousness—the one from the want of the substance; the other from the want of the form. But the persistence of this element under successive conditions, *necessitutes* a sense of it as distinguished from the conditions, and independent of them. The sense of a something that is conditioned in every thought, cannot be got rid of, because the something cannot be got rid of. How then must the sense of this something be constituted? Evidently by combining successive concepts deprived of their limits and conditions. We form this indefinite thought, as we form many of our definite thoughts, by the coalescence of a series of thoughts. Let me illustrate this. A large complex object, having attributes too numerous to be represented at once, is yet tolerably well conceived by the union of several representations, each standing for part of its attributes. On thinking of a piano, there first rises in imagination its visual appearance, to which are instantly added (though by separate mental acts) the ideas of its remote side and of its solid substance. A complete conception, however, involves the strings, the hammers, the dampers, the pedals; and while successively adding these to the conception, the attributes first thought of lapse more or less completely out of consciousness. Nevertheless, the whole group constitutes a representation of the piano. Now as in this case we form a definite concept of a special existence, by imposing limits and conditions in successive acts; so, in the converse case, by taking away the limits and conditions in successive acts, we form an indefinite notion of general existence. By fusing a series of states of consciousness, in each of which, as it arises, the limitations and conditions are abolished, there is produced a consciousness of something unconditioned. To speak more rigorously:—this consciousness is not the abstract of any one group of thoughts, ideas, or conceptions; but it is the abstract of *all* thoughts, ideas, or conceptions. That which is common to them all, and cannot be got rid of, is what we predicate by the word existence. Dissociated as this becomes from each of its modes by the perpetual change of those modes, it remains as an indefinite consciousness of something constant under all modes—of being apart from its appearances. The distinction we feel between special and general existence, is the distinction between that which is changeable in us, and that which is unchangeable. The contrast between the Absolute and the Relative in our minds, is really the contrast between that mental element which exists absolutely, and those which exist relatively.

By its very nature, therefore, this ultimate mental element is at once necessarily indefinite and necessarily indestructible. Our consciousness of the unconditioned being literally the unconditioned consciousness, or raw material of thought to which in thinking we give definite forms, it follows that an ever-present sense of real existence is the very basis of our intelligence. As we can in successive mental acts get rid of all particular conditions and replace them by others, but cannot get rid of that undifferentiated substance of consciousness which is conditioned anew in every thought; there ever remains with us a sense of that which exists persistently and independently of conditions. At the same time that by the laws of thought we are rigorously prevented from forming a conception of absolute existence; we are by the laws of thought equally prevented from ridding ourselves of the consciousness of absolute existence: this consciousness being, as we here see, the obverse of our self-consciousness. And since the only possible measure of relative validity among our beliefs, is the degree of their persistence in opposition to the efforts made to change them, it follows that this which persists at all times, under all circumstances, and cannot cease until consciousness ceases, has the highest validity of any.

To sum up this somewhat too elaborate argument:—We have seen how in the very assertion that all our knowledge, properly so called, is Relative, there is involved the assertion that there exists a Non-relative. We have seen how, in each step of the argument by which this doctrine is established, the same assumption is made. We have seen how, from the very necessity of thinking in relations, it follows that the Relative is itself inconceivable, except as related to a real Non-relative. We have seen that unless a real Non-relative or Absolute be postulated, the Relative itself becomes absolute; and so brings the argument to a contradiction. And on contemplating the process of thought, we have equally seen how impossible it is to get rid of the consciousness of an actuality lying behind appearances; and how, from this impossibility, results our indestructible belief in that actuality.

CHAPTER V
THE RECONCILIATION

§ 27. Thus do all lines of argument converge to the same conclusion. The inference reached *à priori*. in the last chapter, confirms the inferences which, in the two preceding chapters, were reached *à posteriori*. Those imbecilities of the understanding that disclose themselves when we try to answer the highest questions of objective science, subjective science proves to be necessitated by the laws of that understanding. We not only learn by the frustration of all our efforts, that the reality underlying appearances is totally and for ever inconceivable by us; but we also learn why, from the very nature of our intelligence, it must be so. Finally we discover that this conclusion, which, in its unqualified form, seems opposed to the instinctive convictions of mankind, falls into harmony with them when the missing qualification is supplied. Though the Absolute cannot in any manner or degree be known, in the strict sense of knowing, yet we find that its positive existence is a necessary datum of consciousness; that so long as consciousness continues, we cannot for an instant rid it of this datum; and that thus the belief which this datum constitutes, has a higher warrant than any other whatever.

Here then is that basis of agreement we set out to seek. This conclusion which objective science illustrates, and subjective science shows to be unavoidable,—this conclusion which, while it in the main expresses the doctrine of the English school of philosophy, recognizes also a soul of truth in the doctrine of the antagonist German school—this conclusion which brings the results of speculation into harmony with those of common sense; is also the conclusion which reconciles Religion with Science. Common Sense asserts the existence of a reality; Objective Science proves that this reality cannot be what we think it; Subjective Science shows why we cannot think of it as it is, and yet are compelled to think of it as existing; and in this assertion of a Reality utterly inscrutable in nature, Religion finds an assertion essentially coinciding with her own. We are obliged to regard every phenomenon as a manifestation of some Power by which we are acted upon; phenomena being, so far as we can ascertain, unlimited in their diffusion, we are obliged to regard this Power as omnipresent; and criticism teaches us that this Power is wholly incomprehensible. In this consciousness of an Incomprehensible Omnipresent Power, we have just that consciousness on which Religion dwells. And so we arrive at the point where Religion and Science coalesce.

To understand fully how real is the reconciliation thus reached, it will be needful to look at the respective attitudes that Religion and Science have all along maintained towards this conclusion. We must observe how, all along, the imperfections of each have been undergoing correction by the other; and how the final out-come of their mutual criticisms, can be nothing else than an entire agreement on this deepest and widest of all truths.

§ 28. In Religion let us recognize the high merit that from the beginning it has dimly discerned the ultimate verity, and has never ceased to insist upon it. In its earliest and crudest forms it manifested, however vaguely and inconsistently, an intuition forming the germ of this highest belief in which all philosophies finally unite. The consciousness of a mystery is traceable in the rudest fetishism. Each higher religious creed, rejecting those definite and simple interpretations of Nature previously given, has become more religious by doing this. As the quite concrete and conceivable agencies alleged as the causes of things, have been replaced by agencies less concrete and conceivable, the element of mystery has of necessity become more predominant. Through all its successive phases the disappearance of those positive dogmas by which the mystery was made unmysterious, has formed the essential change delineated in religious history. And so Religion has ever been approximating towards that complete recognition of this mystery which is its goal.

For its essentially valid belief, Religion has constantly done battle. Gross as were the disguises under which it first espoused this belief, and cherishing this belief, though it still does, under disfiguring vestments, it has never ceased to maintain and defend it. It has everywhere established and propagated one or other modification of the doctrine that all things are manifestations of a Power that transcends our knowledge. Though from age to age, Science has continually defeated it wherever they have come in collision, and has obliged it to relinquish one or more of its positions; it has still held the remaining ones with undiminished tenacity. No exposure of the logical inconsistency of its conclusions—no proof that each of its particular dogmas was absurd, has been able to weaken its allegiance to that ultimate verity for which it stands. After criticism has abolished all its arguments and reduced it to silence, there has still remained with it the indestructible consciousness of a truth which, however faulty the mode in which it had been expressed, was yet a truth beyond cavil. To this conviction its adherence has been substantially sincere. And for the guardianship and diffusion of it, Humanity has ever been, and must ever be, its debtor.

But while from the beginning, Religion has had the all-essential office of preventing men from being wholly absorbed in the relative or immediate, and of awakening them to a consciousness of something beyond it, this office has been but very imperfectly discharged. Religion has ever been more or less irreligious; and it continues to be partially irreligious even now. In the first place, as implied above, it has all along professed to have some knowledge of that which transcends knowledge; and has so contradicted its own teachings. While with one breath it has asserted that the Cause of all things passes understanding, it has, with the next breath, asserted that the Cause of all things possesses such or such attributes—can be in so far understood. In the second place, while in great part sincere in its fealty to the great truth it had had to uphold, it has often been insincere, and consequently irreligious, in maintaining the untenable doctrines by which it has obscured this great truth. Each assertion respecting the nature, acts, or motives of that Power which the Universe manifests to us, has been repeatedly called in question, and proved to be inconsistent with itself, or with accompanying assertions. Yet each of them has been age after age insisted on, in spite of a secret consciousness that it would not bear examination. Just as though unaware that its central position was impregnable, Religion has obstinately held every outpost long after it was obviously indefensible. And this naturally introduces us to the third and most serious form of irreligion which Religion has displayed; namely, an imperfect belief in that which it especially professes to believe. How truly its central position *is* impregnable, Religion has never adequately realized. In the devoutest faith as we habitually see it, there lies hidden an innermost core of scepticism; and it is this scepticism which causes that dread of inquiry displayed by Religion when face to face with Science. Obliged to abandon one by one the superstitions it once tenaciously held, and daily finding its cherished beliefs more and more shaken, Religion shows a secret fear that all things may some day be explained; and thus itself betrays a lurking doubt whether that Incomprehensible Cause of which it is conscious, is really incomprehensible.

Of Religion then, we must always remember, that amid its many errors and corruptions it has asserted and diffused a supreme verity. From the first, the recognition of this supreme verity, in however imperfect a manner, has been its vital element; and its various defects, once extreme but gradually diminishing, have been so many failures to recognize in full that which it recognized in part. The truly religious element of Religion has always been good; that which has proved untenable in doctrine and vicious in practice, has been its irreligious element; and from this it has been ever undergoing purification.

§ 29. And now observe that all along, the agent which has effected the purification has been Science. We habitually overlook the fact that this has been one of its functions. Religion ignores its immense debt to Science; and Science is scarcely at all conscious how much Religion owes it. Yet it is demonstrable that every step by which Religion has progressed from its first low conception to the comparatively high one it has now reached, Science has helped it, or rather forced it, to take; and that even now, Science is urging further steps in the same direction.

Using the word Science in its true sense, as comprehending all positive and definite knowledge of the order existing among surrounding phenomena, it becomes manifest that from the outset, the discovery of an established order has modified that conception of disorder, or undetermined order, which underlies every superstition. As fast as experience proves that certain familiar changes always happen in the same sequence, there begins to fade from the mind the conception of a special personality to whose variable will they were before ascribed. And when, step by step, accumulating observations do the like with the less familiar changes, a similar modification of belief takes place with respect to them.

While this process seems to those who effect, and those who undergo it, an anti-religious one, it is really the reverse. Instead of the specific comprehensible agency before assigned, there is substituted a less specific and less comprehensible agency; and though this, standing in opposition to the previous one, cannot at first call forth the same feeling, yet, as being less comprehensible, it must eventually call forth this feeling more fully. Take an instance. Of old the Sun was regarded as the chariot of a god, drawn by horses. How far the idea thus grossly expressed, was idealized, we need not inquire. It suffices to remark that this accounting for the apparent motion of the Sun by an agency like certain visible terrestrial agencies, reduced a daily wonder to the level of the commonest intellect. When, many centuries after, Kepler discovered that the planets moved round the Sun in ellipses and described equal areas in equal times, he concluded that in each planet there must exist a spirit to guide its movements. Here we see that with the progress of Science, there had disappeared the idea of a gross mechanical traction, such as was first assigned in the case of the Sun; but that while for this there was substituted an indefinite and less-easily conceivable force, it was still thought needful to assume a special personal agent as a cause of the regular irregularity of motion. When, finally, it was proved that these planetary revolutions with all their variations and disturbances, conformed to one universal law—when the presiding spirits which Kepler conceived were set aside, and the force of gravitation put in their place; the change was really the abolition of an imaginable agency, and the substitution of an unimaginable one. For though the *law* of gravitation is within our mental grasp, it is impossible to realize in thought the *force* of gravitation. Newton himself confessed the force of gravitation to be incomprehensible without the intermediation of an ether; and, as we have already seen, (§ 18,) the assumption of an ether does not in the least help us. Thus it is with Science in general. Its progress in grouping particular relations of phenomena under laws, and these special laws under laws more and more general, is of necessity a progress to causes that are more and more abstract. And causes more and more abstract, are of necessity causes less and less conceivable; since the formation of an abstract conception involves the dropping of certain concrete elements of thought. Hence the most abstract conception, to which Science is ever slowly approaching, is one that merges into the inconceivable or unthinkable, by the dropping of all concrete elements of thought. And so is justified the assertion, that the beliefs which Science has forced upon Religion, have been intrinsically more religious than those which they supplanted.

Science however, like Religion, has but very incompletely fulfilled its office. As Religion has fallen short of its function in so far as it has been irreligious; so has Science fallen short of its function in so far as it has been unscientific. Let us note the several parallelisms. In its earlier stages, Science, while it began to teach the constant relations of phenomena, and so discredited the belief in separate personalities as the causes of them, itself substituted the belief in causal agencies which, if not personal, were yet concrete. When certain facts were said to show "Nature's abhorrence of a vacuum," when the properties of gold were explained as due to some entity called "aureity," and when the phenomena of life were attributed to "a vital principle;" there was set up a mode of interpreting the facts, which, while antagonistic to the religious mode, because assigning other agencies, was also unscientific, because it professed to know that about which nothing was known. Having abandoned these metaphysical agencies—having seen that they were not independent existences, but merely special combinations of general causes, Science has more recently ascribed extensive groups of phenomena to electricity, chemical affinity, and other like general powers. But in speaking of these as ultimate and independent entities, Science has preserved substantially the same attitude as before. Accounting thus for all phenomena, those of Life and Thought included, it has not only maintained its seeming antagonism to Religion, by alleging agencies of a radically unlike kind; but, in so far as it has tacitly assumed a knowledge of these agencies, it has continued unscientific. At the present time, however, the most advanced men of science are abandoning these later conceptions, as their predecessors abandoned the earlier ones. Magnetism, heat, light &c,

which were awhile since spoken of as so many distinct imponderables, physicists are now beginning to regard as different modes of manifestation of some one universal force; and in so doing are ceasing to think of this force as comprehensible. In each phase of its progress, Science has thus stopped short with superficial solutions—has unscientifically neglected to ask what was the nature of the agents it so familiarly invoked. Though in each succeeding phase it has gone a little deeper, and merged its supposed agents in more general and abstract ones, it has still, as before, rested content with these as if they were ascertained realities. And this, which has all along been the unscientific characteristic of Science, has all along been a part cause of its conflict with Religion.

§ 30. We see then that from the first, the faults of both Religion and Science have been the faults of imperfect development. Originally a mere rudiment, each has been growing into a more complete form; the vice of each has in all times been its incompleteness; the disagreements between them have throughout been nothing more than the consequences of their incompleteness; and as they reach their final forms, they come into entire harmony.

The progress of intelligence has throughout been dual. Though it has not seemed so to those who made it, every step in advance has been a step towards both the natural and the supernatural. The better interpretation of each phenomenon has been, on the one hand, the rejection of a cause that was relatively conceivable in its nature but unknown in the order of its actions, and, on the other hand, the adoption of a cause that was known in the order of its actions but relatively inconceivable in its nature. The first advance out of universal fetishism, manifestly involved the conception of agencies less assimilable to the familiar agencies of men and animals, and therefore less understood; while, at the same time, such newly-conceived agencies in so far as they were distinguished by their uniform effects, were better understood than those they replaced. All subsequent advances display the same double result. Every deeper and more general power arrived at as a cause of phenomena, has been at once less comprehensible than the special ones it superseded, in the sense of being less definitely representable in thought; while it has been more comprehensible in the sense that its actions have been more completely predicable. The progress has thus been as much towards the establishment of a positively unknown as towards the establishment of a positively known. Though as knowledge approaches its culmination, every unaccountable and seemingly supernatural fact, is brought into the category of facts that are accountable or natural; yet, at the same time, all accountable or natural facts are proved to be in their ultimate genesis unaccountable and supernatural. And so there arise two antithetical states of mind, answering to the opposite sides of that existence about which we think. While our consciousness of Nature under the one aspect constitutes Science, our consciousness of it under the other aspect constitutes Religion.

Otherwise contemplating the facts, we may say that Religion and Science have been undergoing a slow differentiation; and that their ceaseless conflicts have been due to the imperfect separation of their spheres and functions. Religion has, from the first, struggled to unite more or less science with its nescience; Science has, from the first, kept hold of more or less nescience as though it were a part of science. Each has been obliged gradually to relinquish that territory which it wrongly claimed, while it has gained from the other that to which it had a right; and the antagonism between them has been an inevitable accompaniment of this process. A more specific statement will make this clear. Religion, though at the outset it asserted a mystery, also made numerous definite assertions respecting this mystery—professed to know its nature in the minutest detail; and in so far as it claimed positive knowledge, it trespassed upon the province of Science. From the times of early mythologies, when such intimate acquaintance with the mystery was alleged, down to our own days, when but a few abstract and vague propositions are maintained, Religion has been compelled by Science to give up one after another of its dogmas—of those assumed cognitions which it could not substantiate. In the mean time, Science substituted for the personalities to which Religion ascribed phenomena, certain metaphysical entities; and in doing this it trespassed on the province of Religion; since it classed among the things which it comprehended, certain forms of the incomprehensible. Partly by the criticisms of Religion, which has occasionally called in question its assumptions, and partly as a consequence of spontaneous growth, Science has been obliged to abandon these attempts to include within the boundaries of knowledge that which cannot be known; and has so yielded up to Religion that which of right belonged to it. So long as this process of differentiation is incomplete, more or less of antagonism must continue. Gradually as the limits of possible cognition are established, the causes of conflict will diminish. And a permanent peace will be reached when Science becomes fully convinced that its explanations are proximate and relative; while Religion becomes fully convinced that the mystery it contemplates is ultimate and absolute.

Religion and Science are therefore necessary correlatives. As already hinted, they stand respectively for those two antithetical modes of consciousness which cannot exist asunder. A known cannot be thought of apart from an unknown; nor can an unknown be thought of apart from a known. And by consequence neither can become more distinct without giving greater distinctness to the other. To carry further a metaphor before used, — they are the positive and negative poles of thought; of which neither can gain in intensity without increasing the intensity of the other.

§ 31. Thus the consciousness of an Inscrutable Power manifested to us through all phenomena, has been growing ever clearer; and must eventually be freed from its imperfections. The certainty that on the one hand such a Power exists, while on the other hand its nature transcends intuition and is beyond imagination, is the certainty towards which intelligence has from the first been progressing. To this conclusion Science inevitably arrives as it reaches its confines; while to this conclusion Religion is irresistibly driven by criticism. And satisfying as it does the demands of the most rigorous logic at the same time that it gives the religious sentiment the widest possible sphere of action, it is the conclusion we are bound to accept without reserve or qualification.

Some do indeed allege that though the Ultimate Cause of things cannot really be thought of by us as having specified attributes, it is yet incumbent upon us to assert these attributes. Though the forms of our consciousness are such that the Absolute cannot in any manner or degree be brought within them, we are nevertheless told that we must represent the Absolute to ourselves under these forms. As writes Mr Mansel, in the work from which I have already quoted largely — "It is our duty, then, to think of God as personal; and it is our duty to believe that He is infinite."

That this is not the conclusion here adopted, needs hardly be said. If there be any meaning in the foregoing arguments, duty requires us neither to affirm nor deny personality. Our duty is to submit ourselves with all humility to the established limits of our intelligence; and not perversely to rebel against them. Let those who can, believe that there is eternal war set between our intellectual faculties and our moral obligations. I for one, admit no such radical vice in the constitution of things.

This which to most will seem an essentially irreligious position, is an essentially religious one — nay is *the* religious one, to which, as already shown, all others are but approximations. In the estimate it implies of the Ultimate Cause, it does not fall short of the alternative position, but exceeds it. Those who espouse this alternative position, make the erroneous assumption that the choice is between personality and something lower than personality; whereas the choice is rather between personality and something higher. Is it not just possible that there is a mode of being as much transcending Intelligence and Will, as these transcend mechanical motion? It is true that we are totally unable to conceive any such higher mode of being. But this is not a reason for questioning its existence; it is rather the reverse. Have we not seen how utterly incompetent our minds are to form even an approach to a conception of that which underlies all phenomena? Is it not proved that this incompetency is the incompetency of the Conditioned to grasp the Unconditioned? Does it not follow that the Ultimate Cause cannot in any respect be conceived by us because it is in every respect greater than can be conceived? And may we not therefore rightly refrain from assigning to it any attributes whatever, on the ground that such attributes, derived as they must be from our own natures, are not elevations but degradations? Indeed it seems somewhat strange that men should suppose the highest worship to lie in assimilating the object of their worship to themselves. Not in asserting a transcendant difference, but in asserting a certain likeness, consists the element of their creed which they think essential. It is true that from the time when the rudest savages imagined the causes of all things to be creatures of flesh and blood like themselves, down to our own time, the degree of assumed likeness has been diminishing. But though a bodily form and substance similar to that of man, has long since ceased, among cultivated races, to be a literally-conceived attribute of the Ultimate Cause—though the grosser human desires have been also rejected as unfit elements of the conception—though there is some hesitation in ascribing even the higher human feelings, save in greatly idealized shapes; yet it is still thought not only proper, but imperative, to ascribe the most abstract qualities of our nature. To think of the Creative Power as in all respects anthropomorphous, is now considered impious by men who yet hold themselves bound to think of the Creative Power as in some respects anthropomorphous; and who do not see that the one proceeding is but an evanescent form of the other. And then, most marvellous of all, this course is persisted in even by those who contend that we are wholly unable to frame any conception whatever of the Creative Power. After it has been shown that every supposition respecting the genesis of the Universe commits us to alternative impossibilities of thought—after it has been shown that each attempt to conceive real existence ends in an intellectual suicide—after it has been shown why, by the very constitution of our minds, we are eternally debarred from thinking of the Absolute; it is still asserted that we ought to think of the Absolute thus and thus. In all imaginable ways we find thrust upon us the truth, that we are not permitted to know—nay are not even permitted to conceive—that Reality which is behind the veil of Appearance; and yet it is said to be our duty to believe (and in so far to conceive) that this Reality exists in a certain defined manner. Shall we call this reverence? or shall we call it the reverse?

Volumes might be written upon the impiety of the pious. Through the printed and spoken thoughts of religious teachers, may almost everywhere be traced a professed familiarity with the ultimate mystery of things, which, to say the least of it, seems anything but congruous with the accompanying expressions of humility. And surprisingly enough, those tenets which most clearly display this familiarity, are those insisted upon as forming the vital elements of religious belief. The attitude thus assumed, can be fitly represented only by further developing a simile long current in theological controversies—the simile of the watch. If for a moment we made the grotesque supposition that the tickings and other movements of a watch constituted a kind of consciousness; and that a watch possessed of such a consciousness, insisted on regarding the watchmaker's actions as determined like its own by springs and escapements; we should simply complete a parallel of which religious teachers think much. And were we to suppose that a watch not only formulated the cause of its existence in these mechanical terms, but held that watches were bound out of reverence so to formulate this cause, and even vituperated, as atheistic watches, any that did not venture so to formulate it; we should merely illustrate the presumption of theologians by carrying their own argument a step further. A few extracts will bring home to the reader the justice of this comparison. We are told, for example, by one of high repute among religious thinkers, that the Universe is "the manifestation and abode of a Free Mind, like our own; embodying His personal thought in its adjustments, realizing His own ideal in its phenomena, just as we express own inner faculty and character through the natural language of an external life. In this view, we interpret Nature by Humanity; we find the key to her aspects in such purposes and affections as our own consciousness enables us to conceive; we look everywhere for physical signals of an ever-living Will; and decipher the universe as the autobiography of an Infinite Spirit, repeating itself in miniature within our Finite Spirit." The same writer goes still further. He not only thus parallels the assimilation of the watchmaker to the watch,—he not only thinks the created can "decipher" "the autobiography" of the Creating; but he asserts that the necessary limits of the one are necessary limits of the other. The primary qualities of bodies, he says, "belong eternally to the material datum objective to God" and control his acts; while the secondary ones are "products of pure Inventive Reason and Determining Will"—constitute "the realm of Divine originality." * * * "While on this Secondary field His Mind and ours are thus contrasted, they meet in resemblance again upon the Primary: for the evolutions of deductive Reason there is but one track possible to all intelligences; no *merum arbitrium* can interchange the false and true, or make more than one geometry, one scheme of pure Physics, for all worlds; and the Omnipotent Architect Himself, in realizing the Kosmical conception, in shaping the orbits out of immensity and determining seasons out of eternity, could but follow the laws of curvature, measure and proportion." That is to say, the Ultimate Cause is like a human mechanic, not only as "shaping" the "material datum objective to" Him, but also as being obliged to conform to the necessary properties of that "datum." Nor is this all. There follows some account of "the Divine psychology," to the extent of saying that "we learn" "the character of God—the order of affections in Him" from "the distribution of authority in the hierarchy of our impulses." In other words, it is alleged that the Ultimate Cause has desires that are to be classed as higher and lower like our own.[7] Every one has heard of the king who wished he had been present at the creation of the world, that he might have given good advice. He was humble however compared with those who profess to understand not only the relation of the Creating to the created, but also how the Creating is constituted. And yet this transcendant audacity, which claims to penetrate the secrets of the Power manifested to us through all existence—nay even to stand behind that Power and note the conditions to its action—this it is which passes current as piety! May we not without hesitation affirm that a sincere recognition of the truth that our own and all other existence is a mystery absolutely and for ever beyond our comprehension, contains more of true religion than all the dogmatic theology ever written?

Meanwhile let us recognize whatever of permanent good there is in these persistent attempts to frame conceptions of that which cannot be conceived. From the beginning it has been only through the successive failures of such conceptions to satisfy the mind, that higher and higher ones have been gradually reached; and doubtless, the conceptions now current are indispensable as transitional modes of thought. Even more than this may be willingly conceded. It is possible, nay probable, that under their most abstract forms, ideas of this order will always continue to occupy the background of our consciousness. Very likely there will ever remain a need to give shape to that indefinite sense of an Ultimate Existence, which forms the basis of our intelligence. We shall always be under the necessity of contemplating it as *some* mode of being; that is—of representing it to ourselves in *some* form of thought, however vague. And we shall not err in doing this so long as we treat every notion we thus frame as merely a symbol, utterly without resemblance to that for which it stands. Perhaps the constant formation of such symbols and constant rejection of them as inadequate, may be hereafter, as it has hitherto been, a means of discipline. Perpetually to construct ideas requiring the utmost stretch of our faculties, and perpetually to find that such ideas must be abandoned as futile imaginations, may realize to us more fully than any other course, the greatness of that which we vainly strive to grasp. Such efforts and failures may serve to maintain in our minds a due sense of the incommensurable difference between the Conditioned and the Unconditioned. By continually seeking to know and being continually thrown back with a deepened conviction of the impossibility of knowing, we may keep alive the consciousness that it is alike our highest wisdom and our highest duty to regard that through which all things exist as The Unknowable.

§ 32. An immense majority will refuse with more or less of indignation, a belief seeming to them so shadowy and indefinite. Having always embodied the Ultimate Cause so far as was needful to its mental realization, they must necessarily resent the substitution of an Ultimate Cause which cannot be mentally realized at all. "You offer us," they say, "an unthinkable abstraction in place of a Being towards whom we may entertain definite feelings. Though we are told that the Absolute is real, yet since we are not allowed to conceive it, it might as well be a pure negation. Instead of a Power which we can regard as having some sympathy with us, you would have us contemplate a Power to which no emotion whatever can be ascribed. And so we are to be deprived of the very substance of our faith."

This kind of protest of necessity accompanies every change from a lower creed to a higher. The belief in a community of nature between himself and the object of his worship, has always been to man a satisfactory one; and he has always accepted with reluctance those successively less concrete conceptions which have been forced upon him. Doubtless, in all times and places, it has consoled the barbarian to think of his deities as so exactly like himself in nature, that they could be bribed by offerings of food; and the assurance that deities could not be so propitiated, must have been repugnant, because it deprived him of an easy method of gaining supernatural protection. To the Greeks it was manifestly a source of comfort that on occasions of difficulty they could obtain, through oracles, the advice of their gods, — nay, might even get the personal aid of their gods in battle; and it was probably a very genuine anger which they visited upon philosophers who called in question these gross ideas of their mythology. A religion which teaches the Hindoo that it is impossible to purchase eternal happiness by placing himself under the wheel of Juggernaut, can scarcely fail to seem a cruel one to him; since it deprives him of the pleasurable consciousness that he can at will exchange miseries for joys. Nor is it less clear that to our Catholic ancestors, the beliefs that crimes could be compounded for by the building of churches, that their own punishments and those of their relatives could be abridged by the saying of masses, and that divine aid or forgiveness might be gained through the intercession of saints, were highly solacing ones; and that Protestantism, in substituting the conception of a God so comparatively unlike ourselves as not to be influenced by such methods, must have appeared to them hard and cold. Naturally, therefore, we must expect a further step in the same direction to meet with a similar resistance from outraged sentiments. No mental revolution can be accomplished without more or less of laceration. Be it a change of habit or a change of conviction, it must, if the habit or conviction be strong, do violence to some of the feelings; and these must of course oppose it. For long-experienced, and therefore definite, sources of satisfaction, have to be substituted sources of satisfaction that have not been experienced, and are therefore indefinite. That which is relatively well known and real, has to be given up for that which is relatively unknown and ideal. And of course such an exchange cannot be made without a conflict involving pain. Especially then must there arise a strong antagonism to any alteration in so deep and vital a conception as that with which we are here dealing. Underlying, as this conception does, all others, a modification of it threatens to reduce the superstructure to ruins. Or to change the metaphor—being the root with which are connected our ideas of goodness, rectitude, or duty, it appears impossible that it should be transformed without causing these to wither away and die. The whole higher part of the nature almost of necessity takes up arms against a change which, by destroying the established associations of thought, seems to eradicate morality.

This is by no means all that has to be said for such protests. There is a much deeper meaning in them. They do not simply express the natural repugnance to a revolution of belief, here made specially intense by the vital importance of the belief to be revolutionized; but they also express an instinctive adhesion to a belief that is in one sense the best—the best for those who thus cling to it, though not abstractedly the best. For here let me remark that what were above spoken of as the imperfections of Religion, at first great but gradually diminishing, have been imperfections only as measured by an absolute standard; and not as measured by a relative one. Speaking generally, the religion current in each age and among each people, has been as near an approximation to the truth as it was then and there possible for men to receive: the more or less concrete forms in which it has embodied the truth, have simply been the means of making thinkable what would otherwise have been unthinkable; and so have for the time being served to increase its impressiveness. If we consider the conditions of the case, we shall find this to be an unavoidable conclusion. During each stage of evolution, men must think in such terms of thought as they possess. While all the conspicuous changes of which they can observe the origins, have men and animals as antecedents, they are unable to think of antecedents in general under any other shapes; and hence creative agencies are of necessity conceived by them in these shapes. If during this phase, these concrete conceptions were taken from them, and the attempt made to give them comparatively abstract conceptions, the result would be to leave their minds with none at all; since the substituted ones could not be mentally represented. Similarly with every successive stage of religious belief, down to the last. Though, as accumulating experiences slowly modify the earliest ideas of causal personalities, there grow up more general and vague ideas of them; yet these cannot be at once replaced by others still more general and vague. Further experiences must supply the needful further abstractions, before the mental void left by the destruction of such inferior ideas can be filled by ideas of a superior order. And at the present time, the refusal to abandon a relatively concrete notion for a relatively abstract one, implies the inability to frame the relatively abstract one; and so proves that the change would be premature and injurious. Still more clearly shall we see the injuriousness of any such premature change, on observing that the effects of a belief upon conduct must be diminished in proportion as the vividness with which it is realized becomes less. Evils and benefits akin to those which the savage has personally felt, or learned from those who have felt them, are the only evils and benefits he can understand; and these must be looked for as coming in ways, like those of which he has had experience. His deities must be imagined to have like motives and passions and methods with the beings around him; for motives and passions and methods of a higher character, being unknown to him, and in great measure unthinkable by him, cannot be so realized in thought as to influence his deeds. During every phase of civilization, the actions of the Unseen Reality, as well as the resulting rewards and punishments, being conceivable only in such forms as experience furnishes, to supplant them by higher ones before wider experiences have made higher ones conceivable, is to set up vague and uninfluential motives for definite and influential ones. Even now, for the great mass of men, unable through lack of culture to trace out with due clearness those good and bad consequences which conduct brings round through the established order of the Unknowable, it is needful that there should be vividly depicted future torments and future joys—pains and pleasures of a definite kind, produced in a manner direct and simple enough to be clearly imagined. Nay still more must be conceded. Few if any are as yet fitted wholly to dispense with such conceptions as are current. The highest abstractions take so great a mental power to realize with any vividness, and are so inoperative upon conduct unless they are vividly realized, that their regulative effects must for a long period to come be appreciable on but a small minority. To see clearly how a right or wrong act generates consequences, internal and external, that go on branching out more widely as years progress, requires a rare power of analysis. To mentally represent even a single series of these consequences, as it stretches out into the remote future, requires an equally rare power of imagination. And to estimate these consequences in their totality, ever multiplying in number while diminishing in intensity, requires a grasp of thought possessed by none. Yet it is only by such analysis, such imagination, and such grasp, that conduct can be rightly guided in the absence of all other control: only so can ultimate rewards and penalties be made to outweigh proximate pains and pleasures. Indeed, were it not that throughout the progress of the race, men's experiences of the effects of conduct have been slowly generalized into principles—were it not that these principles have been from generation to generation insisted on by parents, upheld by public opinion, sanctified by religion, and enforced by threats of eternal damnation for disobedience— were it not that under these potent influences, habits have been modified, and the feelings proper to them made innate—were it not, in short, that we have been rendered in a considerable degree organically moral; it is certain that disastrous results would ensue from the removal of those strong and distinct motives which the current belief supplies. Even as it is, those who relinquish the faith in which they have been brought up, for this most abstract faith in which Science and Religion unite, may not uncommonly fail to act up to their convictions. Left to their organic morality, enforced only by general reasonings imperfectly wrought out and difficult to keep before the mind, their defects of nature will often come out more strongly than they would have done under their previous creed. The substituted creed can become adequately operative only when it becomes, like the present one, an element in early education, and has the support of a strong social sanction. Nor will men be quite ready for it until, through the continuance of a discipline which has already partially moulded them to the conditions of social existence, they are completely moulded to those conditions.

We must therefore recognize the resistance to a change of theological opinion, as in great measure salutary. It is not simply that strong and deep-rooted feelings are necessarily excited to antagonism—it is not simply that the highest moral sentiments join in the condemnation of a change which seems to undermine their authority; but it is that a real adaptation exists between an established belief and the natures of those who defend it; and that the tenacity of the defence measures the completeness of the adaptation. Forms of religion, like forms of government, must be fit for those who live under them; and in the one case as in the other, that form which is fittest is that for which there is an instinctive preference. As certainly as a barbarous race needs a harsh terrestrial rule, and habitually shows attachment to a despotism capable of the necessary rigour; so certainly does such a race need a belief in a celestial rule that is similarly harsh, and habitually shows attachment to such a belief. And just in the same way that the sudden substitution of free institutions for tyrannical ones, is sure to be followed by a reaction; so, if a creed full of dreadful ideal penalties is all at once replaced by one presenting ideal penalties that are comparatively gentle, there will inevitably be a return to some modification of the old belief. The parallelism holds yet further. During those early stages in which there is an extreme incongruity between the relatively best and the absolutely best, both political and religious changes, when at rare intervals they occur, are necessarily violent; and necessarily entail violent retrogressions. But as the incongruity between that which is and that which should be, diminishes, the changes become more moderate, and are succeeded by more moderate retrogressions; until, as these movements and counter-movements decrease in amount and increase in frequency, they merge into an almost continuous growth. That adhesion to old institutions and beliefs, which, in primitive societies, opposes an iron barrier to any advance, and which, after the barrier has been at length burst through, brings back the institutions and beliefs from that too-forward position to which the momentum of change had carried them, and so helps to re-adapt social conditions to the popular character—this adhesion to old institution and beliefs, eventually becomes the constant check by which the constant advance is prevented from being too rapid. This holds true of religious creeds and forms, as of civil ones. And so we learn that theological conservatism, like political conservatism, has an all-important function.

§ 33. That spirit of toleration which is so marked a characteristic of modern times, and is daily growing more conspicuous, has thus a far deeper meaning than is supposed. What we commonly regard simply as a due respect for the right of private judgment, is really a necessary condition to the balancing of the progressive and conservative tendencies—is a means of maintaining the adaptation between men's beliefs and their natures. It is therefore a spirit to be fostered; and it is a spirit which the catholic thinker, who perceives the functions of these various conflicting creeds, should above all other men display. Doubtless whoever feels the greatness of the error to which his fellows cling and the greatness of the truth which they reject, will find it hard to show a due patience. It is hard for him to listen calmly to the futile arguments used in support of irrational doctrines, and to the misrepresentation of antagonist doctrines. It is hard for him to bear the manifestation of that pride of ignorance which so far exceeds the pride of science. Naturally enough such a one will be indignant when charged with irreligion because he declines to accept the carpenter-theory of creation as the most worthy one. He may think it needless as it is difficult, to conceal his repugnance to a creed which tacitly ascribes to The Unknowable a love of adulation such as would be despised in a human being. Convinced as he is that all punishment, as we see it wrought out in the order of nature, is but a disguised beneficence, there will perhaps escape from him an angry condemnation of the belief that punishment is a divine vengeance, and that divine vengeance is eternal. He may be tempted to show his contempt when he is told that actions instigated by an unselfish sympathy or by a pure love of rectitude, are intrinsically sinful; and that conduct is truly good only when it is due to a faith whose openly-professed motive is other-worldliness. But he must restrain such feelings. Though he may be unable to do this during the excitement of controversy, or when otherwise brought face to face with current superstitions, he must yet qualify his antagonism in calmer moments; so that his mature judgment and resulting conduct may be without bias.

To this end let him ever bear in mind three cardinal facts—two of them already dwelt upon, and one still to be pointed out. The first is that with which we set out; namely the existence of a fundamental verity under all forms of religion, however degraded. In each of them there is a soul of truth. Through the gross body of dogmas traditions and rites which contain it, it is always visible—dimly or clearly as the case may be. This it is which gives vitality even to the rudest creed; this it is which survives every modification; and this it is which we must not forget when condemning the forms under which it is presented. The second of these cardinal facts, set forth at length in the foregoing section, is, that while those concrete elements in which each creed embodies this soul of truth, are bad as measured by an absolute standard, they are good as measured by a relative standard. Though from higher perceptions they hide the abstract verity within them; yet to lower perceptions they render this verity more appreciable than it would otherwise be. They serve to make real and influential over men, that which would else be unreal and uninfluential. Or we may call them the protective envelopes, without which the contained truth would die. The remaining cardinal fact is, that these various beliefs are parts of the constituted order of things; and not accidental but necessary parts. Seeing how one or other of them is everywhere present; is of perennial growth; and when cut down, redevelops in a form but slightly modified; we cannot avoid the inference that they are needful accompaniments of human life, severally fitted to the societies in which they are indigenous. From the highest point of view, we must recognize them as elements in that great evolution of which the beginning and end are beyond our knowledge or conception—as modes of manifestation of The Unknowable; and as having this for their warrant.

Our toleration therefore should be the widest possible. Or rather, we should aim at something beyond toleration, as commonly understood. In dealing with alien beliefs, our endeavour must be, not simply to refrain from injustice of word or deed; but also to do justice by an open recognition of positive worth. We must qualify our disagreement with as much as may be of sympathy.

§ 34. These admissions will perhaps be held to imply, that the current theology should be passively accepted; or, at any rate, should not be actively opposed. "Why," it may be asked, "if all creeds have an average fitness to their times and places, should we not rest content with that to which we are born? If the established belief contains an essential truth—if the forms under which it presents this truth, though intrinsically bad, are extrinsically good—if the abolition of these forms would be at present detrimental to the great majority—nay, if there are scarcely any to whom the ultimate and most abstract belief can furnish an adequate rule of life; surely it is wrong, for the present at least, to propagate this ultimate and most abstract belief."

The reply is, that though existing religious ideas and institutions have an average adaptation to the characters of the people who live under them; yet, as these characters are ever changing, the adaptation is ever becoming imperfect; and the ideas and institutions need remodelling with a frequency proportionate to the rapidity of the change. Hence, while it is requisite that free play should be given to conservative thought and action, progressive thought and action must also have free play. Without the agency of both, there cannot be those continual re-adaptations which orderly progress demands.

Whoever hesitates to utter that which he thinks the highest truth, lest it should be too much in advance of the time, may reassure himself by looking at his acts from an impersonal point of view. Let him duly realize the fact that opinion is the agency through which character adapts external arrangements to itself—that his opinion rightly forms part of this agency—is a unit of force, constituting, with other such units, the general power which works out social changes; and he will perceive that he may properly give full utterance to his innermost conviction: leaving it to produce what effect it may. It is not for nothing that he has in him these sympathies with some principles and repugnance to others. He, with all his capacities, and aspirations, and beliefs, is not an accident, but a product of the time. He must remember that while he is a descendant of the past, he is a parent of the future; and that his thoughts are as children born to him, which he may not carelessly let die. He, like every other man, may properly consider himself as one of the myriad agencies through whom works the Unknown Cause; and when the Unknown Cause produces in him a certain belief, he is thereby authorized to profess and act out that belief. For, to render in their highest sense the words of the poet—

——Nature is made better by no mean,

But nature makes that mean: over that art

Which you say adds to nature, is an art

That nature makes.

Not as adventitious therefore will the wise man regard the faith which is in him. The highest truth he sees he will fearlessly utter; knowing that, let what may come of it, he is thus playing his right part in the world—knowing that if he can effect the change he aims at—well: if not—well also; though not *so* well.

7. These extracts are from an article entitled "Nature and God," published in the *National Review* for October, 1860.